WANTING

WANTING

THE POWER OF

MIMETIC DESIRE

IN EVERYDAY LIFE

Luke Burgis

ST. MARTIN'S PRESS

NEW YORK

First published in the United States by St. Martin's Press, an imprint of St. Martin's Publishing Group

WANTING. Copyright © 2021 by Luke Burgis. All rights reserved. Printed in the United States of America. For information, address St. Martin's Publishing Group, 120 Broadway, New York, NY 10271.

Jenny Holzer quote, page 99, © 2020 Jenny Holzer, member Artists Rights Society (ARS), New York

www.stmartins.com

Design by Meryl Sussman Levavi

Illustrations by Liana Finck

Library of Congress Cataloging-in-Publication Data

Names: Burgis, Luke, author.
Title: Wanting : the power of mimetic desire in everyday life / Luke Burgis.
Description: First edition. | New York : St. Martin's Press, [2021] |
    Includes bibliographical references and index.
Identifiers: LCCN 2020057564 | ISBN 9781250262486 (hardcover) |
    ISBN 9781250262493 (ebook)
Subjects: LCSH: Desire. | Imitation. | Basic needs—Psychological aspects.
Classification: LCC BF575.D4 B87 2021 | DDC 150.19/5—dc23
LC record available at https://lccn.loc.gov/2020057564

Our books may be purchased in bulk for promotional, educational, or business use. Please contact your local bookseller or the Macmillan Corporate and Premium Sales Department at 1-800-221-7945, extension 5442, or by email at MacmillanSpecialMarkets@macmillan.com.

First Edition: 2021

10 9 8 7 6 5 4 3 2 1

To Claire and Hope

CONTENTS

TACTICS

Imitation is natural to man from childhood, one of his advantages over the lower animals being this, that he is the most imitative creature in the world.

—Aristotle

We want what other people want because other people want it, and it's penciled-in eyebrows all the way down, down to the depths of the nth circle of hell where we all die immediately of a Brazilian butt lift, over and over again.

—Dayna Tortorici

NOTE TO READER

This is a book about why people want what they want. Why *you* want what you want.

Each of us spends every moment of our life, from the moment we're born to the moment we die, wanting something. We even want in our sleep. Yet few people ever take the time to understand how they come to want things in the first place.

Wanting well, like thinking clearly, is not an ability we're born with. It's a freedom we have to earn. Due to one powerful yet little-known feature of human desire, that freedom is hard-won.

I spent my twenties starting companies, chasing the entrepreneurial dream that Silicon Valley had enshrined in me. I was searching for financial freedom, I thought, and the recognition and respect that come with it.

Then something odd happened: when I walked away from one of the companies that I had founded, I experienced intense relief.

That's when I realized I hadn't *found* anything at all. My previous successes had felt like failures, and now failure felt like success. What was the force behind my tenacious and never-satisfied striving?

This crisis of meaning led me to spend a lot of time in libraries and bars. Sometimes I brought the library to the bar. (I'm not kidding. Once I brought a backpack full of books to a sports bar during the World Series and tried to read with Phillies fans celebrating all around me.) I traveled to Thailand and Tahiti. I worked out like a maniac.

But it all seemed like palliative care, not the treatment of the underlying condition. While this period helped me think more seriously about my choices, it didn't help me understand the desires that had led me to those choices in the first place—the navigation system behind my ambition.

One day a mentor suggested I look into a set of ideas that would, he told me, explain why I had come to want the things I wanted, and how my desires entrapped me in cycles of passion followed by disillusionment.

The source of those ideas was a fairly obscure but influential academic. Before he died on November 4, 2015, at ninety-one, René Girard was named an immortal of the Académie Française and called "the new Darwin of the social sciences." As a professor at Stanford in the 1980s through the mid-1990s, he inspired a small group of followers. Some of them believed that his ideas would be the key to understanding the twenty-first century—and that when the history of the twentieth century is written, circa 2100, he would be seen as the most important thinker of his generation.[1]

And you have probably never heard of him.

René Girard's mind drew other people to him from every direction. To begin with, he had an uncanny ability to notice things that explained mysterious human behavior. He was a Sherlock Holmes of history and literature, putting his finger on overlooked clues while everyone else was busy following the usual suspects.

He was playing a different game than other academics. He was like the only person at a poker table who has identified the tell of the dominant player. While other players are calculating the mathematical odds of having the winning hand, he's staring into faces. He's watching his rival to see how many times he blinks and whether he picks the cuticle of his left index finger.

Girard identified a fundamental truth about desire that connected the seemingly unconnected: linking biblical stories with volatility in the stock market, the collapse of ancient civilizations with workplace dysfunction, career paths with diet trends. He explained, well before they existed, why Facebook, Instagram, and their progeny have been so wildly popular and effective in selling people both stuff and dreams.

Girard discovered that most of what we desire is mimetic (mi-**met**-ik) or imitative, not intrinsic. Humans *learn*—through imitation—to *want* the same things other people want, just as they learn how to speak the same language and play by the same cultural rules. Imitation plays a far more pervasive role in our society than anyone had ever openly acknowledged.

Our power of imitation dwarfs that of any other animal. It allows us to build sophisticated culture and technology. At the same time, it has a dark side. Imitation leads people to pursue things that seem desirable at first but ultimately leave them unfulfilled. It locks them into cycles of desire and rivalry that are difficult, practically impossible, to escape.

But Girard offered his students hope. It *was* possible to transcend the cycles of frustrated desire. It was possible to have more agency in shaping the life we want.

My introduction to Girard took me from *Oh, shit* to *Holy shit*. Mimetic theory would help me recognize patterns in the behavior of people and in current events. That was the easy part. Later, after seeing mimetic desire everywhere except in my own life, I saw it in myself—my *Holy shit* moment. Mimetic theory would eventually help me uncover and declutter my own messy world of desires. And that process was hard.

I'm now convinced that understanding mimetic desire is the key to understanding, at a deeply human level, business, politics, economics, sports, art, even love. It can help you make money, if that is your primary driver. Or it may help you avoid waiting until middle age or later to learn that money or prestige or a comfortable life is not primarily what you want.

Mimetic theory sheds light on what motivates economic and political and personal tensions, and also shows the way out of them. For those with a creative spirit, it can guide their creativity to projects that create real human and economic value and not just wealth transfers.

I don't claim that overcoming mimetic desire is possible, or even desirable. This book is primarily about growing more aware of its presence so that we can navigate it better. Mimetic desire is like gravity—it just *is*. Gravity is always at work. It causes some people to live in constant pain when they don't develop the muscles in their core and around their spine to be able to stand up straight and face the world, to resist the downward pull. Others experience that same gravity and find ways to go to the moon.

Mimetic desire is like that. If we're not aware of it, it will take us places we don't want to go. But if we develop the right social and emotional muscles in response to it, mimetic desire becomes a a way to effect positive change.

The change you make is up to you—or at least it will be by the end of this book.

There is a growing community of people interested in mimetic theory that spans the political left and right, cuts across siloed disciplines, and stretches across many countries where the divides are different but the theory's explanatory power is the same. The diversity of perspectives suggests that perhaps there is a profound truth about humanity at its core.

Scholars interested in Girard's work have made important contributions on topics ranging from the hermeneutics of mimesis in Shakespeare to the sexual violence against women in war zones to the scapegoating process that happened in the Rwandan genocide. Suffice it to say that people who associate mimetic theory only with Girard's former student, Peter Thiel, and link it with libertarianism or Thiel's politics have an incomplete picture. I wrote this book in part to break up the monopoly that he has in some people's minds as the interpreter of Girard's thought. He would be the first to tell you that's a good thing. Ideological monopolies are the worst monopolies.

Mimetic desire transcends the political. It is in some sense pre-political, kind of like comedy. When something is funny, it's funny. But even humor can become tainted and bound up with agendas and rivalries. If any reader finishes this book and uses insights that might be found in it to attack their enemies, they will have missed a key point.

At a time of rising tension in the United States and many other parts of the world—at least while I was writing this—I wanted to offer something that might encourage more reflection and restraint, a recognition of our rivalries, and a hope that we can live with neighbors who want different things than we do.

These days I spend part of my time mentoring aspiring entrepreneurs. Their ambition to build a better world and live full, meaningful lives inspires me. But I worry that if they don't understand how desire works, they'll wind up disappointed.

The idea of *being* an entrepreneur has high mimetic value these days. Nearly every budding entrepreneur I know is motivated to achieve some form of freedom. But running your own company does not automatically lead to more freedom. Sometimes it leads to the opposite. We think of entrepreneurs as the ultimate renegades, not bound to a 9-to-5 desk job or serving as a middle-management cog in a sclerotic machine. Yet thinking that you don't have a traditional boss might just mean that mimetic desire is your tyrant. I push my students to look deeper.

I can't guarantee them success in business, but I can guarantee that by the time they leave my class they will not be wanting naively. They will move forward choosing majors, starting companies, finding partners, and reading the news with a better awareness of what's happening inside them. That awareness is the precondition for change.

There are certain insights that, once you see them, begin to seep into your experience of daily living. An understanding of mimetic desire is one of them. Once you know how it works, you will start to see how it explains much of the world around you. And that will include not just the family member whose weird lifestyle you would never choose, or the politics in your workplace, or the friend who cares about social media too much, or the colleague who brags about their kid who got into Harvard. It will include you. You will see it in yourself.

PROLOGUE
Unexpected Relief

In the summer of 2008, I experienced the moment many start-up founders live for: I learned that I would be able to cash out on my company's success. After an intense period of courtship spanning several months, I was on my way to have celebratory drinks with the CEO of Zappos, Tony Hsieh. Zappos was going to acquire my e-commerce company for wellness products, FitFuel.com.

About an hour earlier, Tony had sent me a direct message on Twitter (his preferred form of communication at the time) asking me to meet him at the Foundation Room, a bar on the sixty-third floor of the Mandalay Bay Hotel in Las Vegas. I knew that he had attended a board meeting earlier that day and that one of the agenda items was the acquisition. He wouldn't be inviting me to the Strip if the news was not good.

I'd been pacing around my house all day. I needed the deal to go through. Fit Fuel was burning through cash. Despite our rapid growth over the past two years, the coming months looked ominous. The Federal Reserve had gone into bailout mode and held an emergency meeting to keep the giant investment bank Bear Stearns from going under. The housing market was crashing. I needed to raise a round of investment, but investors were spooked. All of them told me to come back in a year—but I didn't have a year.

Neither Tony nor I knew at the time how volatile 2008 would turn out to be. At the start of the year, Zappos had exceeded its operating profit goals and decided to award all of its employees generous bonuses. By the end of the year—only eight months after the bonuses were doled out—Zappos would have to lay off 8 percent of its workforce. Already that summer,

Zappos's board members and experienced investors, led by Sequoia Capital, were tightening their belts.[1]

When I got the invitation from Tony, I sped from my home in Henderson, Nevada, to the Strip, blasting old-school hip-hop and letting out intermittent yelps of relief and excitement through the sunroof so that by the time I got there I might seem calm.

In those days Zappos was a nine-year-old company that had recently surpassed $1 billion in sales. Tony conducted unorthodox social experiments, such as offering new hires up to $2,000 to leave the company after their orientation (the idea was that this would separate out employees who were not passionate enough about working there from those who were). The company was well known for its idiosyncratic culture.

Culture is what Tony seemed to like best about Fit Fuel. When he and the other Zappos top brass came to visit our offices and warehouse, they told me how much they liked what they saw: we were scrappy (because understaffed), zany (because everyone at the company was a character), and just the right amount of weird (because we had the trappings of a start-up, like hookah pipes and beanbag chairs).

Tony told me that he wanted me to run the operation as a new division within Zappos. I would build the company's next billion-dollar vertical. Shoes had been the first. Wellness would be the second.

In addition to life-changing money and Zappos equity, I'd be part of a respected leadership team and get paid a nice salary. (I hadn't drawn a regular salary from my companies, ever, and I craved that stability.)

I was not what anyone would call a Zappos "culture fit." But since we were talking about joining forces, I started conforming myself more to the mold of the Zappos culture to make things work.

In my desperation to sell the company, I told Tony everything that I thought he wanted to hear. I had heterodox views about the Zappos culture, out of step with the media's portrayal—but I buried them. It's easy to be an armchair contrarian. It's hard to take contrarian action: to question the dominant narrative, to be honest with yourself, to tell the truth even when the immediate outcome is pain—like losing the chance to sell my company, and instead getting buried under an avalanche of debt.

I try to have skin in the game. This time I had too much.[2]

I'd spent the last few months getting to know Tony. We met after I sent him a cold email and he invited me to lunch at Claim Jumper, a restaurant near the Zappos HQ in Henderson, a suburb of Las Vegas. When I showed up for what I thought was a casual get-to-know-you lunch with him, at least six senior executives were sitting around the table waiting for me. It was an interview. I never had time to touch my clam chowder.

After lunch, Tony and I walked back to his office together. He stopped along the way and put his hands in his pockets as if he was fumbling for change. "So I wouldn't be doing my job," he said, "if I didn't ask you if you'd be open to joining forces." I said yes, and the next few months seemed like a wild engagement period. I was invited to Zappos happy hours, parties at Tony's house, and early morning hikes up Black Mountain.

Tony didn't look like a guy who had millions. He had sold the first company he co-founded, LinkExchange, to Microsoft for $265 million in 1998 at the age of twenty-four. But he dressed in plain jeans and a Zappos T-shirt and drove a dirty Mazda 6. Within a few weeks of hanging out with him, I ditched my True Religions (I know) and started shopping at the Gap. I began to wonder if I should drive an older and dirtier car.

I had co-founded Fit Fuel about three years before I met Tony, in 2005. We had a grandiose mission statement to make healthier foods more accessible for everyone in the world. I chipped away, day after day, making steady progress and learning how to lead a growing company. But even as our sales increased and accolades rolled in, I experienced a declining desire to go into the office every day.

Tim Ferriss's book *The 4-Hour Workweek: Escape 9–5, Live Anywhere, and Join the New Rich* hit shelves while I was struggling to figure things out. *If I'm working more than four hours a week, I must be doing something wrong*, I thought. I began frantically looking around for better models of entrepreneurship, but I couldn't be sure who was telling the truth.

Meeting Tony only amplified my despair. I was shooting for $10 million in sales. Zappos was doing $1 billion. From my perspective, Tony occupied an alternate reality—the one in which unicorn founders live. I couldn't seem to break in.

I experienced a sort of existential vertigo, like I was jumping off the top of a skyscraper onto a giant trampoline that catapulted me back to the top before I plunged back down again. What I wanted seemed to change daily: more respect and status, less responsibility; more capital, fewer investors; more public speaking, more privacy; an intense lust for money followed by extreme bouts of virtue signaling involving the word *social*. I even vacillated between wanting to bulk up and trying to slim down.

The most troubling thing to me was that the desire that led me to start and build my company was gone. Where did it go? Where had it come from in the first place? My desires felt like rom-com love—things I fell into rather than things I chose. (By the way, did you know that in almost every language in the world, people *fall* in love? Nobody *rises up* into it.[3])

Meanwhile, the internal conflict between my co-founder and me got worse until we agreed to go our separate ways. I took over as sole leader of the company at the very time I had lost the desire to lead.

It was clear that there were mysterious forces outside myself that affected what I wanted and how intensely I wanted it. I couldn't make any serious decisions until I knew more about them. I couldn't start another company. I was even hesitant about the thought of getting married someday, knowing that my desire for something (or someone) one day might be gone the next. Discovering what those forces were seemed like a responsibility.

The day after my celebration drinks with Tony on the Vegas Strip, I took a friend on a tour of the Zappos headquarters, excited to show him my future home. As we walked by Monkey Row (Zappos jargon for the place where the executives sit), I noticed that the execs' faces looked like they'd seen a ghost. We had an awkward exchange.

It was the bad feeling before a breakup.

My friend and I went out for dinner later that night. In the middle of our pastas I received a call from Alfred Lin, who between 2005 and 2010 was the CFO, COO, and chairman of Zappos.

Alfred sounded somber. Then he told me why.

After the official board meeting, the Zappos board of directors had had a second meeting on the plane back to San Francisco and decided to put any immediate plans on hold. There would be no acquisition. "They changed their minds," he said.

"They *changed their minds*?" I asked.

"Yeah. I don't know what more to say," Alfred said. "I'm sorry."

"They *changed their minds*?" I kept asking the same question, and Alfred kept telling me the same thing. I kept mouthing the words after I hung up the phone, but this time as a statement, not a question. "They changed . . . their minds." I repeated it as I walked back to the table, sat down, and stared into my bowl of bad spaghetti, prodding and twirling it endlessly, making perfect bites only to unravel them and start all over again.

There would be no life-changing exit, no windfall, no second home in Sicily. Worse, my company was on the rocks. Without the Zappos deal, I'd be bankrupt within six months. As the full import of how my life was about to change sank in and I drained my Chianti, something changed.

I was relieved.

WANTING

INTRODUCTION
Social Gravity

On the far wall hangs a photograph—a single black-and-white eyeball looking out, cropped close, no bigger than a coaster, matted in a twenty-two-inch frame.

I'm sitting in the home of Peter Thiel above the Sunset Strip. Thiel is known for being the billionaire co-founder of PayPal, for being the first outside investor in Facebook, for his contrarian views on business, and for taking down Gawker and publicly challenging Google. But I'm not here to talk to him about any of those things.

A few minutes pass, and the assistant who showed me in returns to check on me. "Peter will be with you shortly. Anything else I can bring you, sir? More coffee?"

"Oh, no, thanks," I say. I'm embarrassed that I've chugged my entire cup. He smiles and exits.

This two-story living room would be at home in any midcentury *Architectural Digest* spread. Floor-to-ceiling paneled windows open onto an infinity pool overlooking Sunset Boulevard. It's homey but still grandiose.

The focal piece of the spacious room is a wet bar built into an oak-paneled gallery wall featuring artwork in cool hues: black-and-white photos, deep indigo prints, gray etchings. Among them are an inkblot, maybe a Rorschach, shaped like a crab; a large print that contains abstract circles and rods, possibly molecular geometry; and a triptych of a man standing waist-deep in what looks like icy mountain lake water.

Elsewhere in the room, starker elements are set off by soft velvet couches and armchairs. In the center of the six-inch-thick wood coffee table in front of me, a silver teardrop-shaped metallic sculpture balances defiantly on its point. Twenty-foot-high double doors—the likes of which I've only seen in cathedrals—lead into the next room. Near the door is a chess table waiting for a worthy challenger. (It won't be me.) A telescope points out a window next to a Greek bust. Everything hangs together. If the movie *Clue* had been directed by Ray Eames, it would look like Peter Thiel's house.

A man appears on a second-story exposed walkway on the far side of the room. "Be with you in a minute," Peter Thiel says.

He waves his hand and smiles, then disappears through a door. I hear running water. Ten minutes later, he reemerges in a baseball tee, shorts, and running shoes. He descends the spiral staircase.

"Hi, I'm Peter," he says, extending his hand. "So you're here to talk about Girard's ideas."

A Dangerous Mind

René Girard, a Frenchman who was a professor of literature and history in the United States, had his first insight about the nature of desire in the late 1950s. It would change his life. Three decades later, when Peter Thiel was an undergraduate philosophy major at Stanford, the professor would alter his life, too.

The discovery that changed Girard's life in the 1950s and Thiel's in the 1980s (and mine in the 2000s) is mimetic desire. It's what's brought me to Thiel's home. I was drawn to mimetic theory, quite simply, because I'm mimetic. We all are.

Mimetic theory isn't like learning some impersonal law of physics, which you can study from a distance. It means learning something new about your own past that explains how your identity has been shaped and why certain people and things have exerted more influence over you than others. It means coming to grips with a force that permeates human relationships— relationships which you are, at this moment, involved in. You can never be a neutral observer of mimetic desire.

Thiel and I have both experienced the disconcerting moment when we discovered that force at work in our lives. It's so personal that I hesitated to write a book about it. To write about mimetic desire is to reveal a bit of your own.

I ask him why he didn't explicitly mention Girard in his popular business book *Zero to One*, even though it was packed with insights from his mentor.[1] "There's something dangerous about Girard's ideas," Thiel says. "I think people have self-defense mechanisms against some of this stuff." He wanted people to see that Girard's insights contain important truths and that they explain what is going on in the world around them, but he didn't want to take his readers all the way through the looking-glass.

An idea that challenges commonly held assumptions can feel threatening—and that's all the more reason to look more closely at it: to understand why.

An unbelieved truth is often more dangerous than a lie. The lie in this case is the idea that I want things entirely on my own, uninfluenced by others, that I'm the sovereign king of deciding what is wantable and what is not. The truth is that my desires are derivative, mediated by others, and that I'm part of an ecology of desire that is bigger than I can fully understand.

By embracing the lie of my independent desires, I deceive only myself. But by rejecting the truth, I deny the consequences that my desires have for other people and theirs for me.

It turns out the things we want matter far more than we know.

Like Henry Ford seeing the assembly line in a slaughterhouse, or like psychologist Daniel Kahneman shaping the new field of behavioral economics, Girard's breakthrough came when he was outside his main area of expertise, history. It happened when he was forced to apply his thinking to classic novels.

Early in his academic career in the United States, Girard was asked to teach literature courses covering books that he hadn't yet read. Reluctant to turn down work, he agreed. Often he would read the novels in the syllabus just in time to turn around and teach them. He read and taught Cervantes, Stendhal, Flaubert, Dostoyevsky, Proust, and more.

With his lack of formal training and the need to read quickly, he started to look for patterns in the texts. He uncovered something perplexing, something which seemed to be present in nearly all of the most compelling novels ever written: characters in these novels rely on other characters to show them what is worth wanting. They don't spontaneously desire anything. Instead, their desires are formed by interacting with other characters who alter their goals and their behavior—most of all, their desires.

Girard's discovery was like the Newtonian revolution in physics, in which the forces governing the movement of objects can only be understood in a *relational* context. Desire, like gravity, does not reside autonomously in any one thing or person. It lives in the space between them.[2]

The novels Girard taught are not primarily plot-driven or character-driven. They are desire-driven. A character's actions are a reflection of their desires, which are shaped in relationship to the desires of others. The plots unfold according to who is in a mimetic relationship with whom and how their desires interact and play out.

The two characters don't even have to meet for this relationship to happen. Don Quixote, alone in his room, reads about the adventures of the famous knight Amadís de Gaula. He is inflamed with a desire to emulate him and become a knight-errant, wandering the countryside in search of opportunities to prove the virtues of chivalry.

In all of the books Girard taught, desire always involved an imitator and a model. Other readers had not noticed it, or they had overlooked it by discounting the possibility of such a pervasive theme.

Girard's distance from the subject matter, combined with his penetrating intellect, enabled him to recognize the pattern. The characters in the great novels are so realistic because they want things the way that we do—not spontaneously, not out of an inner chamber of authentic desire, not randomly, but through the imitation of someone else: their secret model.

Messing Up Maslow

Girard discovered that we come to desire many things not through biological drives or pure reason, nor as a decree of our illusory and sovereign self, but through imitation.

That idea was unpalatable to me the first time I heard it. Are we all just imitation machines? No. Mimetic desire is only one piece of a comprehen-

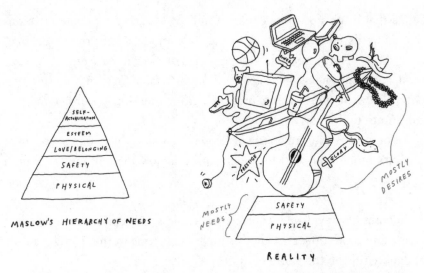

MASLOW'S HIERARCHY OF NEEDS

REALITY

Abraham Maslow's hierarchy of needs is too *neat*. After a person has fulfilled their basic needs, they enter a universe of desires that does not have a stable hierarchy.

sive vision of human ecology, which also includes freedom and a relational understanding of personhood. The imitation of desire has to do with our profound openness to other people's interior lives—something that sets us apart as humans.

Desire, as Girard used the word, does not mean the drive for food or sex or shelter or security. Those things are better called *needs*—they're hardwired into our bodies. Biological needs don't rely on imitation. If I'm dying of thirst in the desert, I don't need anyone to show me that water is desirable.

But after meeting our basic needs as creatures, we enter into the human universe of desire. And knowing what to want is much harder than knowing what to need.

Girard was interested in how we come to want things when there is no clear instinctual basis for it.[3] Out of the billions of potential objects of desire in the world, from friends to careers to lifestyles, how do people come to desire some more than others? And why do the objects and intensity of our desire seem to fluctuate constantly, lacking any real stability?

In the universe of desire, there is no clear hierarchy. People don't choose objects of desire the way they choose to wear a coat in the winter. Instead of internal biological signals, we have a different kind of external signal that motivates these choices: models. Models are people or things that show us what is worth wanting. It is models—not our "objective" analysis or central nervous system—that shape our desires. With these models, people engage in a secret and sophisticated form of imitation that Girard termed *mimesis* (mi-**mee**-sis), from the Greek word *mimesthai* (meaning "to imitate").

Models are the gravitational centers around which our social lives turn. It's more important to understand this now than at any other time in history.

As humans have evolved, people have spent less time concerned about surviving and more time striving for things—less time in the world of needs and more time in the world of desire.

Even water has transitioned from the world of needs to the world of desires. Imagine you came here from another planet that was still in the pre-bottled-water stage of evolution (a critical stage), and I asked you whether you preferred Aquafina, Voss, or San Pellegrino. Which would you choose? Sure, I could present you with the minerality breakdown and pH levels of each, but we'd be kidding ourselves if we think that's how you will make your choice. I tell you I drink San Pellegrino. And if you're an imitative creature like me, or if you just think I'm a more highly developed being than you—because you come from a *pre-Pellegrino* people—you're going to choose San Pellegrino.

If you look hard enough, you will find a model (or a set of models) for almost everything—your personal style, the way you speak, the look and feel of your home. But the models that most of us overlook are models of desire. It's deceivingly difficult to figure out why you bought certain things; it's extraordinarily hard to understand why you strive toward certain achievements. So hard that few people dare to ask.

Mimetic desire draws people toward things.[4] "This draw," writes Girard scholar James Alison, "this movement . . . [is] mimesis. It is to psychology what gravity is to physics."[5]

Gravity causes people to fall physically to the ground. Mimetic desire causes people to fall in or out of love, or debt, or friendships, or business partnerships. Or it may subject them to the degrading slavery of being merely a product of their milieu.

The Evolution of Desire

Back in Peter Thiel's home, Thiel tells me that he's more prone to mimetic behavior than most people. Though he is known by many as a contrarian thinker, he hasn't always been that way.

Like a lot of high school students, he strived to gain admission to a prestigious university (in his case, Stanford) without questioning why he wanted to go there in the first place. It's just what people with his background did.

Once there, the striving continued—for grades, internships, and other totems of success. He noticed that there was a decent degree of diversity among the career goals of newly arrived freshmen. But over the next few years, the goals seemed to converge: finance, law, medicine, or consulting. Thiel had a nagging sense that something was off.

He would gain some insight into the problem when he learned about Professor Girard through a small group of students who were fascinated with his thought. In his junior year, he started attending lunches and gatherings at which he knew the professor would be in attendance.

Girard challenged students to understand both the how and the why behind current events. He could move systematically through human history, showing layer upon layer of meaning, sometimes quoting entire passages of Shakespeare from memory to illustrate his point.

He gave accounts of ancient texts and classic literature with such penetrating insight that his students felt an adrenaline rush, as if they'd set foot in a new universe. One of his earlier students, Sandor Goodhart, now a professor at Purdue University, remembers Girard opening the very first session of his class Literature, Myth, and Prophecy with these words: "Human beings fight not because they are different, but because they are the same, and in their attempts to distinguish themselves have made themselves into enemy twins, human doubles in reciprocal violence."[6] A far cry from the more typical "Okay, welcome to the class, let's go over the syllabus."

After living in France under German occupation during World War II, Girard came to the United States in September 1947 to teach French and work toward a PhD in history at Indiana University. He stood out on the campus in Bloomington: he had a big head and big ideas, and he could be intimidating to the uninitiated.

Girard met his future wife there, an American from Indiana named Martha McCullough. He couldn't pronounce her last name during roll call. They met again about a year later, when Martha was no longer his student. They eventually married.[7]

Girard failed to get tenure at Indiana because he didn't publish enough of his work. He was dismissed. He went on to teach at Duke University, Bryn Mawr College, Johns Hopkins University, and SUNY Buffalo. Finally, in 1981, he became the Andrew B. Hammond Professor of French Language, Literature, and Civilization at Stanford University, where he stayed for the rest of his career until his official retirement in 1995.[8]

To many students and faculty at Stanford, Girard exuded Old World charisma. Cynthia Haven, a writer and scholar long affiliated with the university, remembers a remarkable-looking man with a "totemic" head walking around the campus before she knew who he was. They eventually became friends, and she wrote a biography of him titled *Evolution of Desire: A Life of René Girard*. "He had the sort of face a film director might typecast in a movie to play one of the greatest thinkers of all times," she wrote, "a Plato or a Copernicus."[9]

Girard was an autodidact with wide range. He studied anthropology, philosophy, theology, and literature, integrating them into an original and sophisticated view of the world. He found that mimetic desire was closely related to violence, especially the idea of sacrifice. The biblical story of Cain and Abel is about Cain killing his brother, Abel, after his ritual sacrifice pleased God less than Abel's. They both wanted the same thing—to win favor with God—which brought them into direct conflict with each other. In Girard's view, the root of most violence is mimetic desire.

In a French television show from the 1970s, Girard explains mimetic theory to a panel of interviewers, casually ashing his cigarette as he talks. "What has fascinated me for a very long time is sacrifice," he tells the panelists, "the fact that men, for religious reasons, kill animal and often human victims in almost all human societies."[10] He burned to understand the problem of violence and the religious fascination with sacrifice that extends to nearly every part of human culture.

(Indeed, one of his more controversial claims is that the domestication of cats and dogs could not have come about intentionally. People

René Girard at a SUNY Buffalo arts
faculty meeting, July 1971.
(All photos courtesy of Bruce Jackson.)

Girard at the opening of his seminar in
spring 1971 that would form the basis
of his book *Violence and the Sacred*.

Girard conversing with Diane Christian,
a longtime Distinguished Professor of
English at the University of Buffalo.

Girard in spring 1971 with French literary theorist
Gérard Bucher.

did not originally intend to live with cats and dogs the way we do today, integrating them peacefully into our families for as long as they live. That process would have taken generations of coordinated effort. The reason we domesticated animals was far more practical, he argued: communities integrated the animals into their lives in order to sacrifice them. Sacrifices are more effective when they come from within a community—when the victim has something in common with the sacrificers. We'll discuss why in Chapter 4.[11])

The consequences of mimetic desire play out in strange ways across many different domains. Most of the drama happens behind the scenes.

* * *

Peter Thiel's exposure to Girard did not immediately divert him from the course he was on. He took a job in finance and went to law school. But he felt lost. "I had this core life crisis when I realized that all these hyper-track competitive things I was after were for these bad social reasons," he told me.

Meeting Girard at Stanford had introduced Thiel to the idea of mimesis, but an intellectual understanding didn't immediately translate to changed behavior. "You find yourself trapped in all these bad mimetic cycles," he says. "And there was a lot of resistance—a doctrinaire libertarian resistance—from me. Mimetic theory pushes against the idea that we're all these atomistic individuals." The flattery of self-sufficiency is powerful. "That took me a while to overcome," Thiel says.

He describes a transformation that was both intellectual and existential. Once he learned about mimetic desire, he could identify it when he saw it—in everybody but himself.

"The intellectual transformation was quick because that was something I was looking for," he says. But he continued to struggle after graduation because he didn't see the extent to which he was embroiled in the very things Girard had been talking about. "The existential dimension took me a while to percolate though."

Thiel left the corporate world and co-founded Confinity with Max Levchin in 1998. He began to use his knowledge of mimetic theory to help him manage both the business and his life. When competitive rivalries flared up within his company, he gave each employee clearly defined and independent tasks so they didn't compete with one another for the same responsibilities. This is important in a start-up environment where roles are often fluid. A company in which people are evaluated based on clear performance objectives—not their performance relative to one another—minimizes mimetic rivalries.

When there was risk of an all-out war with Elon Musk's rival company, X.com, Thiel merged with him to form PayPal. He knew from Girard that when two people (or two companies) take each other as mimetic models, they enter into a rivalry for which there is no end but destruction—unless they are somehow able to see beyond the rivalry.[12]

Thiel took mimesis into account when evaluating investment decisions, too. Reid Hoffman, the founder of LinkedIn, had introduced him to Mark Zuckerberg. Thiel saw clearly that Facebook was not merely another MySpace or SocialNet (Hoffman's first start-up). Facebook was built around identity—that is to say, desires. It helps people see what other people have and want. It is a platform for finding, following, and differentiating oneself from models.

Models of desire are what make Facebook such a potent drug. Before Facebook, a person's models came from a small set of people: friends, family, work, magazines, and maybe TV. After Facebook, everyone in the world is a potential model.

Facebook isn't filled with just *any* kind of model—most people we follow aren't movie stars, pro athletes, or celebrities. Facebook is full of models who are *inside our world*, socially speaking. They are close enough for us to compare ourselves to them. They are the most influential models of all, and there are billions of them.

Thiel quickly grasped Facebook's potential power and became its first outside investor. "I bet on mimesis," he told me. His $500,000 investment eventually yielded him over $1 billion.

What's at Stake

Mimetic desire, because it is social, spreads from person to person and through a culture. It results in two different movements—two cycles—of desire. The first cycle leads to tension, conflict, and volatility, breaking down relationships and causing instability and confusion as competing desires interact in volatile ways. This is the default cycle that has been most prevalent in human history. It is accelerating today.

It's possible to transcend that default cycle, though. It's possible to initiate a different cycle that channels energy into creative and productive pursuits that serve the common good.

This book will explore these two cycles. They're fundamental to human behavior. Because they are so close to us—because they operate *within* us—we tend to look past them. Yet these cycles are at work constantly.

Movements of desire are what define our world. Economists measure them, politicians poll them, businesses feed them. History is the story of human desire. Yet the origin and evolution of desire are mysterious. Girard titled his 1978 magnum opus *Things Hidden Since the Foundation of the World*. It was a nod to the lengths that humans have gone to hide the true nature of their desires and their consequences. This book is about those hidden things and how they play out in the world today. We can't afford to ignore them because:

1. **Mimesis can hijack our noblest ambitions.**

   We live at a time of hyper-imitation. Fascination with what is trending and going viral is symptomatic of our predicament. So is political polarization. It stems in part from mimetic behavior that destroys nuance and poisons even our most honorable goals: to develop friendships, to fight for important causes, to build healthy communities. When mimesis takes over, we become obsessed with vanquishing some Other, and we measure ourselves according to them. When a person's identity becomes completely tied to a mimetic model, they can never truly escape that model because doing so would mean destroying their own reason for being.[13]

2. **Homogenizing forces are creating a crisis of desire.**

   Equality is good. Sameness is generally not—unless we're talking about cars on an assembly line or the consistency of your favorite brand of coffee. The more that people are forced to be the same—the more pressure they feel to think and feel and want the same things—the more intensely they fight to differentiate themselves. And this is dangerous. Many cultures have had a myth in which twins commit violence against each other. There are at least five separate stories of sibling rivalry in the book of Genesis alone: Cain and Abel, Ishmael and Isaac, Esau and Jacob, Leah and Rachel, Joseph and his brothers. Stories of sibling rivalry are universal because they're true—the more people are alike, the more likely they are to feel threatened. While technology is bringing the world closer together (Facebook's stated mission), it is bringing our desires closer together and amplifying conflict. We are free to resist, but the mimetic forces are accelerating so quickly that we are close to becoming shackled.

3. **Sustainability depends on desirability.**

   Decades of consumer culture have forged unsustainable desires. Many people know *intellectually* that they could do a better job taking care of the planet, for instance. But until eating a more sustainable diet or driving more fuel-efficient cars is far more attractive to the average consumer than the alternatives, the more sustainable options will not be widely adopted. It's not enough to know what is good and true. Goodness and truth need to be attractive—in other words, desirable.

4. **If people don't find positive outlets for their desires, they will find destructive ones.**

In the days before the terrorist attacks of September 11, 2001, hijacker Mohammed Atta and his companions were carousing in south Florida bars and binge-playing video games. "Who asks about the souls of these men?" wondered Girard in his last book, *Battling to the End*.[14] The Manichean division of the world into "evil" and "not evil" people never satisfied him. He saw the dynamics of mimetic rivalry at work in the rise of terrorism and class conflict. People don't fight because they want *different* things; they fight because mimetic desire causes them to want the *same* things. The terrorists would not have been driven to destroy symbols of the West's wealth and culture if, at some deep level, they had not secretly desired some of the same things. That's why the Florida bars and video game–playing are an important piece of the puzzle. The *mysterium iniquitatis* (the mystery of evil) remains just that: mysterious. But mimetic theory reveals something important about it. The more people fight, the more they come to resemble each other. We should choose our enemies wisely, because we become like them.

But even more is at stake. Each one of us has a responsibility to shape the desires of others, just as they shape ours. Each encounter we have with another person enables them, and us, to want more, to want less, or to want differently.

In the final analysis, two questions are critical. *What do you want? What have you helped others want?* One question helps answer the other.

And if you're not satisfied with the answers you find today, that's okay. The most important questions concern what we will want tomorrow.

What Will You Want Tomorrow?

By the end of this book you will have a new understanding of desire—what you want, what others want, and how to live and lead from a model in which desire is an expression of love. To help you get there, this book is a two-part journey.

Part I, "The Power of Mimetic Desire," is about the hidden forces that influence why people want the things they want. It's Mimetic Theory 101. In Chapter 1, I'll start by explaining the origins of mimetic desire in infancy and show how it evolves into a sophisticated form of adult imitation. In Chapter 2, we'll see how mimetic desire works differently depending on a person's relationship with a model. Starting in Chapter 3, I'll explain how mimetic desire works in groups, which is key to understanding some of our most persistent and perplexing societal conflicts. In Chapter 4, we reach the culmination of mimetic conflict: the scapegoat mechanism. The first half of the book focuses on the destructive, or default, cycle of desire: Cycle 1.

Part II, "The Transformation of Desire," outlines a process for breaking free from Cycle 1 in order to manage our desires in a healthier way. The second half of the book shows how we have the freedom to put in motion a creative cycle of desire: Cycle 2. In Chapter 5, we meet a three-Michelin-star chef who stepped outside the system of desire that he was born into and recovered the freedom to create. Chapter 6 shows how disruptive empathy breaks the bonds that keep most of us from discovering and building *thick desires* that make for a good life. Chapter 7 applies the laws of desire to leadership. Finally, Chapter 8 is about the future of desire.

Part I feels like a descent. It's necessary to visit hell so we never become permanent residents. Part II is the way out.

Throughout this book I'll highlight fifteen tactics that I've developed to deal positively with mimetic desire. My goal in sharing them is to help you think practically about these ideas and ultimately to develop your own tactics, which may be very different from mine.

Mimetic desire is part of the human condition. It can lurk under the surface of our lives, acting as our unrecognized leader. But there are ways to recognize it, confront it, and make more intentional choices that lead to a more satisfying life—far more satisfying than one in which we're totally consumed by mimetic desire without knowing it.

By the end of this book, you'll have a simple framework for understanding how desire works in your life and in our culture. You'll have a better idea of *what* you're imitating and *how* you're imitating it. Knowing

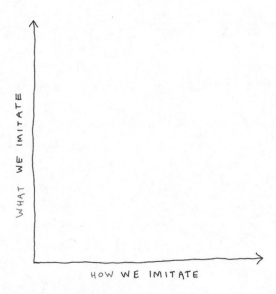

whether you respond more or less mimetically in a given situation and in specific relationships is an important step toward self-mastery.

We're becoming more aware of how fragile and interconnected the world's systems are. Political and economic systems that once seemed stable have been shaken. Public health has been challenged because even the best policies have to contend with groups of people who want different things. The poverty that endures alongside mega-wealth is a scandal. All of these things have a basis in the fundamental system of desire that I'm attempting to describe. This system of desire is to the world's organs what the circulatory system is to the body. When the cardiovascular system isn't working properly, organs suffer and eventually shut down. The same is true of desire.

Our fractured relationships with other humans and with the entire ecosystem reveal that what we want, individually and collectively, has consequences. If we understand the mimetic nature of desire, though, we can play our part in building a better world. The greatest developments in history are the result of someone wanting something that did not yet exist—and helping others to want more than they thought was wantable.

Your new or deepened awareness of mimetic desire will make you see the world differently. If you're like me, it will haunt you to the point that you start seeing it everywhere—maybe even in your own life. What you choose to do about it is up to you.

Part I

THE POWER OF
MIMETIC DESIRE

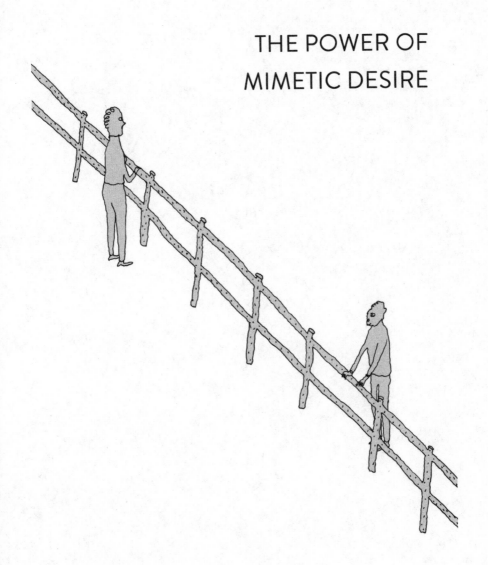

Chapter 1

HIDDEN MODELS
Romantic Lies, Infant Truth

Caesar's Self-Deception . . . Love by Another's Eyes . . .
The Invention of PR . . . Why Playing Hard to Get Works

> We can never know what to want, because, living only one life, we can neither compare it with our previous lives nor perfect it in our lives to come.
>
> —Milan Kundera

When people tell you what they want, they tell a version of the Romantic Lie. It goes something like this:

I just realized that I want to run a marathon. (Like all my friends when they turn thirty-five.)

'Cause I saw a tiger, now I understand . . . (From the song "I Saw a Tiger," written by Vince Johnson for Joe Exotic, the Tiger King, for whom seeing a tiger seems to have been a mystical experience that made him want to start a big-cat zoo.)

I want Christian Grey. I want him badly. Simple fact. (From *Fifty Shades of Grey,* which is full of these simplicities.)

Julius Caesar was an excellent Romantic liar. After his victory at the battle of Zela, he declared, "Veni, vidi, vici" (I came, I saw, I conquered). The line has been quoted thousands of times by people taking Caesar at his word: that he saw the place and decided to conquer it. Magician James Warren suggests that we reframe Caesar's words in the language of desire so we see what he's truly claiming: *I came, I saw, I desired.* And therefore he conquered.[1]

Caesar wants us to think that he needs only to lay eyes on something to know whether it's desirable. But Caesar flatters himself.

The truth is more complex. First, Caesar revered Alexander the Great, the Macedonian military genius who conquered nearly all of the known world in the third century BCE. Second, at the battle of Zela, Caesar's rival, Pharnaces II, had attacked Caesar first. Caesar didn't just *come and see.* He had long desired to conquer like his model, Alexander, and he was responding to his rival, Pharnaces.

The Romantic Lie is self-delusion, the story people tell about why they make certain choices: because it fits their personal preferences, or because they see its objective qualities, or because they simply *saw* it and therefore *wanted* it.

They believe that there is a straight line between them and the things they want. That's a lie. The truth is that the line is always curved.

LIE

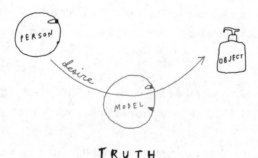

TRUTH

Buried in a deeper layer of our psychology is the person or thing that caused us to want something in the first place. Desire requires models—people who endow things with value for us merely because they want the things.

Models transfigure objects before our eyes. You walk into a consignment store with a friend and see racks filled with hundreds of shirts. Nothing jumps out at you. But the moment your friend becomes enamored with one specific shirt, it's no longer a shirt on a rack. It's *the* shirt that your friend Molly chose—the Molly who, by the way, is an assistant costume designer on major films. The moment she starts ogling the shirt, she sets it apart. It's a different shirt than it was five seconds ago, before she started wanting it.

"O hell! to choose love by another's eyes!" says Hermia in Shakespeare's *A Midsummer Night's Dream*. It's hell to know we have chosen *anything* by another's eyes. But we do it all the time: we choose brands, schools, and dishes at a restaurant by them.

There are always models of desire. If you don't know yours, they are probably wreaking havoc in your life.

You may be wondering, then: if desire is generated and shaped by models, then where do models get their desires? The answer: from other models.

If you go back far enough in the evolution of your desire, through friends and parents and grandparents and great-grandparents, all the way back to the Romans, who modeled themselves after the Greeks, you will keep finding models.

The Bible contains a story about the Romantic Lie at the dawn of humanity. Eve originally had no desire to eat the fruit from the forbidden tree—until the serpent modeled it. The serpent *suggested a desire*. That's what models do. Suddenly, a fruit that had not aroused any particular desire became the most desirable fruit in the universe. Instantaneously. The fruit appeared irresistible because—and only after—it was modeled as a forbidden good.[2]

We are tantalized by models who suggest a desire for things that we don't currently have, especially things that appear just out of reach. The greater the obstacle, the greater the attraction.

Isn't that curious? We don't want things that are too easily possessed or that are readily within reach. Desire leads us beyond where we currently are. Models are like people standing a hundred yards up the road who can see something around the corner that we can't yet see. So the way that a model describes something or *suggests* something to us makes all the difference. We never see the things we want directly; we see them indirectly, like refracted light. We are attracted to things when they are modeled to us in an attractive way, by the right model. Our universe of desire is as big or as small as our models.

Dependence on models isn't necessarily a bad thing. Without models, none of us would be speaking the same language or aspiring to anything other than the status quo. George Carlin might have spent fifty years telling jokes about the weather had he not been in the audience at a 1962 Lenny Bruce performance. Bruce modeled a new way of doing comedy, and Carlin used it to break out.

The danger is not recognizing models for what they are. When we don't recognize them, we are easily drawn into unhealthy relationships with them. They begin to exert an outsize influence on us. We often become fixated on them without realizing it. Models are, in many cases, a person's secret idol.

"René could remove an idol from another person's eye like it was an act of reverence," Girard's friend Gil Bailie told me. Mimetic theory exposes our models and reorders our relationship with them. The first step is bringing them to light.

In this chapter, we'll see how an early twentieth-century hacker named Edward Bernays, "the father of public relations," made use of carefully placed and hidden models to manipulate a generation of consumers. In the 1950s and '60s, his descendants were the "Mad Men" of Madison Avenue. Today, they are more likely to be embedded in large tech companies, governments, and newsrooms.

We'll also see how mimesis affects financial markets, and why finding and naming hidden models helps us understand the movement of stocks and the humanity of bubbles.

But we'll start by looking someplace where models operate in the open: the lives of babies.

Secrets Babies Keep

Babies are brilliant imitators. Mere seconds after birth, they start imitating other humans. As newborns, they're capable of imitation to a degree that surpasses even the most highly developed adult primates.[3]

Researchers have found that a baby's imitative powers start developing even before birth. "After they are born, young babies mimic many different sounds. But they are especially shaped by the prosody they heard in the womb," wrote Sophie Hardach in a 2019 *New York Times* article detailing new research by German scientist Kathleen Wermke. By the third trimester, babies can hear the tones of their mother's voice. Shortly after birth, babies born to Mandarin-speaking mothers (Mandarin is a highly tonal language) tend to cry with more complex intonations than babies born to German- or Swedish-speaking mothers, for example.[4]

These and other recent discoveries have challenged the theory of the asocial infant—the view held by Freud, Skinner, and Piaget that newborn babies are like unhatched chicks, cut off from external reality until adults socialize them. Freud even proposed a distinction between a physical birth and a psychological, or interpersonal, birth, implying that a baby is not a full person until socialized.[5] But any mother who has held a newborn in her arms knows that this is false. Babies are born social.

Few scientists have done more to refute the myth of the asocial infant than Andrew Meltzoff, whose work in childhood development, psychology, and neuroscience over the past several decades has lent support to Girard's discovery. Meltzoff's work suggests that we don't learn how to imitate; we are born imitators. Being an imitator is part of what it means to be human.

In one of his best-known experiments, conducted in 1977, he went to a hospital in Seattle (along with his co-researcher, M. Keith Moore) and stuck out his tongue at newborns. While the mean age of babies in this study was thirty-two hours, an infant as young as forty-two minutes mimicked his facial expressions, mapping onto them with surprising accuracy. It was the first time one of these babies had ever seen another human being stick out their tongue or make funny faces, yet she seemed to realize that she was "like" this creature before her—that she had a face, too, and could do things with it.[6]

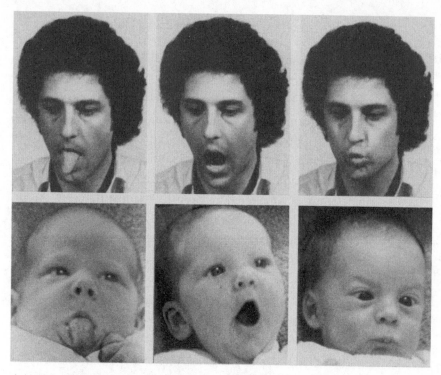

A. N. Meltzoff and M. K. Moore (1977). *Science* 198:75–78. (Photo courtesy of Andrew Meltzoff.)

I traveled to Meltzoff's office at the University of Washington's Institute for Learning and Brain Sciences, which he co-directs along with his wife, Patricia K. Kuhl, a specialist in speech and hearing sciences. "Babies come out of the womb seeming to have the ability to imitate," Meltzoff tells me.

We can understand our mimetic nature by seeing babies as teachers. "Babies hold a secret about the human mind that has been hidden for millennia," Meltzoff wrote. "They are our double. They have a primordial drive to understand us that advances their development; we have a desire to understand them that propels social science and philosophy. By examining the minds and hearts of children, we illuminate ourselves."[7]

Between 2007 and 2009, Meltzoff met René Girard several times at Stanford and once at Girard's house in Palo Alto. They probed one another's insights into the development of human life and culture.

"René was fascinated by the science—the way that infants follow gazes, which draws them into the orbit of adult goals, intentions, and desires," Meltzoff says. Girard appreciated that a scientist of Meltzoff's stature was corroborating and elucidating his theory in a different domain.

"And he suggested some novels I might want to read."

"Novels?" I ask.

"Yes, like Proust."

"What about Proust?"

"He was really interested in a concept in my research called *joint visual attention*: when two people are focused visually on the same object. Babies follow the gaze of their mothers. He pointed me to some wonderful passages in Proust about people paying attention to other people's eyes and reading things about their intentions and desires."

Throughout Proust's masterpiece *In Search of Lost Time*, characters strive to know what others want by paying attention to the slightest signals. "How had I failed to observe long ago," writes Proust in volume 5, "that Albertine's eyes belonged to the category which even in a quite ordinary person seems to be composed of a number of fragments because of all the places in which the person wishes to be—and to conceal the desire to be—on that particular day?"[8] Characters in Proust pick up on cues about what others want, sometimes even from a glance of the eyes.

We do the same thing. Meltzoff explains: "A mother looks at something. A baby takes that as a signal that the mother desires the object, or is at least paying attention to it because it must be important. The baby looks at the mother's face, then at the object. She tries to understand the relationship between her mother and the object." It's not long before a baby can follow not just her mother's eyes but even the intentions behind her actions.

To test that idea, Meltzoff staged an act in front of eighteen-month-old babies. In the experiment, an adult acted as if he was trying to pull apart a dumbbell-shaped toy made from a central tube with a wooden cube on each side. As the adult strained to pull the toy apart, he let his hand slip off one end. He tried again, but this time let his hand slip off the other end. The adult's intention was clear: he wanted to pull the toy apart. But apparently he failed.

After the adult's performance, the researcher presented the object to the infants and observed what they did. The babies would pick up the dumbbell and immediately pull it apart—forty out of the fifty times the experiment was conducted. They didn't mimic what the adults did; they imitated what they thought the adults *wanted to do*. They read beneath the surface behavior.[9]

The babies in the experiment were pre-linguistic. They were tracking the desires of others before they understood or had words to describe them. They didn't know or care about *why* other people wanted something; they simply noticed what they wanted.

Desire is our primordial concern. Long before people can articulate *why* they want something, they start wanting it. The motivational speaker Simon Sinek advises organizations and people to "start with why" (the title of one of his books), finding and communicating one's purpose before anything else. But that is usually a post hoc rationalization of whatever it is we already wanted. Desire is the better place to start.

Children seem surprisingly altruistic. In 2020, Meltzoff and his colleagues observed that nineteen-month-olds will help an adult obtain an out-of-reach piece of fruit. The children in the experiment readily, repeatedly, and rapidly helped others fulfill their desires more than half of the time—and they did this even if it meant handing over a piece of food to an adult when they were hungry.[10]

This natural and healthy concern in children about what other people want seems to morph in adulthood into an *unhealthy* concern about what other people want. It grows into mimesis. Adults do expertly what babies do clumsily. After all, each of us is a highly developed baby. Rather than learning what other people want so that we can help them get it, we secretly compete with them to possess it.

I ask Meltzoff how deep the imitation of desire goes. He jumps from his chair and leads me into a special room that contains a $2 million machine called a magnetoencephalograph (MEG). A MEG has extremely sensitive magnetometers that allow scientists to locate the sources of the magnetic field in the brain. When the brain is active, it produces a magnetic field in and around the head. The machine detects fluctuations in the magnetic fields that someone naturally produces by perceiving, wanting, feeling, and thinking about the things around them.

While the earliest forms of MEG have been around since the 1970s,

Meltzoff's MEG has customized software, including some designed specifically to analyze learning and brain activity in infants. His machine looks
like a hair dryer for giants. It's plastered with stickers of colorful aquatic
creatures.

In a 2018 study, Meltzoff and his team found that a child's brain maps
onto actions they see in the world around them. "We discovered that
when a child sees an adult being touched by an object while the child is
in the MEG, the MEG shows the same part of the brain activate in the
child *as if* the child was being touched themselves."[11] The imaginary divide
between self and other—the foundation of the Romantic Lie—has been
exposed.

Mirror neurons were discovered accidentally, in Parma, Italy, in the 1990s,
by a group of Italian scientists led by prominent neuroscientist Giacomo
Rizzolatti. They discovered that a specific area of a macaque monkey's
brain activates when it sees an adult pick up a peanut—the same area that
lights up if the monkey picks up a peanut directly. "That's why it's called a
mirror neuron," writes Marco Iacoboni, a neuroscientist at the University
of California, Los Angeles. "It's almost like the monkey is watching his own
action reflected by the mirror."[12]

According to Meltzoff, mirror neurons may be one part of the neurological basis for imitation, but they can't explain all of it. "What babies do
is more complicated than mirror neurons," he tells me. What Girard calls
mimetic desire might have some neurological basis in mirror neurons, but
mimetic desire is a mysterious phenomenon that can't be reduced to mirror
neurons alone.

Animals imitate sounds, facial expressions, gestures, aggression, and
other behaviors. Humans imitate all of those things and more: retirement
planning, romantic ideals, sexual fantasies, food preparation, social norms,
worship, gift-giving rituals, professional courtesies, and memes.

We're so sensitive to imitation that we notice the slightest deviance
from what we could call *acceptable imitation*. If we receive a response
to an email or text that doesn't sufficiently tone-match, we can go into a
mini-crisis (*Does she not like me? Does he think he's superior to me? Did I
do something wrong?*). Communication practically runs on mimesis. In a
study published in 2008 in the *Journal of Experimental Social Psychology*,
sixty-two students were assigned to negotiate with other students. Those

who mirrored others' posture and speech reached a settlement 67 percent of the time, while those who didn't reached a settlement 12.5 percent of the time.[13]

The imitation of superficial things is part of everyday life, and usually nothing to worry about—unless it becomes a portal into another world, a black hole in the universe of desire, which can suck us in with no possibility of escape.

The Martini Is a Gateway Drug

The babies whom Meltzoff observed following their mother's gaze became adults who watch their neighbors with rapt attention for the slightest clues about what is desirable.

I plan to order a beer from the bar. My friend orders a gin martini. Suddenly I "realize" that I want a martini, too.

If I'm honest with myself, I *didn't* want a martini when I walked in the door. I had my heart set on a cold beer. What changed? My friend didn't remind me of a subconscious, inner longing that I had for a martini; he gave me a new desire. I want one *because* my friend wanted one first.

A martini is harmless. (Usually.) But let's say that while we're bellied up to the bar sipping our drinks, my friend tells me about a promotion he's about to get. He'll receive a $20,000 boost in salary and have a new title: managing director of something or other that sounds important. It comes with more vacation time, too.

As I smile and tell him how exciting that is, I feel some anxiety. Shouldn't I be making an extra $20,000, too? Will my friend and I still be able to plan vacations together if he gets twice as much paid time off as I do? And also, *what the hell*? We graduated from the same university, and I worked twice as hard as he did in school and after. Am I falling behind? Did I choose the right path in life? Even though I used to say that I could never be in his line of work, now I'm second-guessing myself.

My friend has become a model of desire to me. We will never speak of it. But an inner force has been activated in me that, if left unchecked, will cause conflict. I'll start to make decisions based on what he wants. If he moves to a certain neighborhood, I'll start evaluating where I live accordingly. If he reaches Delta SkyMiles Platinum status, I won't be satisfied with Gold.

Sometimes I'll imitate him in a mirrored way, doing the opposite of whatever he does. If he buys a Tesla, then I'll never want to own a Tesla. I don't want any reminders that I'm always one step behind. I'm different. I'll buy a classic Ford Mustang and start narrowing my eyes at the Tesla drivers I see on the road (*that sheeplike herd . . .*)—completely unaware that my behavior is driven by my model.

When he loses his job, I experience schadenfreude. When he gets it back, I experience envy. Even my emotions are reflections of the relationship I have with my model. And presently I'm at the bottom of my martini glass, and I notice he got an extra olive.

In the passage from childhood to adulthood, the open imitation of the infant becomes the hidden mimesis of adults. We're secretly on the lookout for models while simultaneously denying that we need any.

Mimetic desire operates in the dark. Those who can see in the dark take full advantage.

Torches of Freedom

On April 6, 1917, the same day America declared war on Germany, twenty-five-year-old Edward Bernays began the process of enlisting in the U.S. Army. According to Larry Tye's book about him, Bernays, a five-foot-four Austrian-born Jew who was the nephew of Sigmund Freud, was eager to show his patriotism and defend his adopted country. But flat feet and defective vision disqualified him.[14]

The rejection only motivated Bernays to excel in other ways. He was a sly observer of human nature who had a natural talent for rallying people to causes. He began to wonder how he could exploit that gift.

Four years earlier, as the twenty-one-year-old editor of a small medical reviews magazine, he had seen opportunities where others saw obstacles. Given the magazine's medical focus and the need to generate more interest and readers, Bernays decided to promote the controversial play *Damaged Goods* by the playwright Eugène Brieux. It was about a man with syphilis who gives the disease to his wife, and she goes on to bear a child with the disease. The play had been banned in most places because discussing a sexually transmitted disease was totally taboo at the time. Bernays enlisted medical experts and public role models, including John D. Rockefeller, Anne Harriman Vanderbilt, and Eleanor Roosevelt, to support the play by framing it as a battle against prudishness. Despite mixed reviews, his campaign made the play enormously successful and boosted Bernays's reputation as a shrewd tactician.

He continued to hone his skills with a series of PR maneuvers that tended to connect deeply personal causes to products and entertainment until a far weightier opportunity arose. After being rejected by the military, Bernays set out to use his considerable skills to rally support for America to enter World War I.

The prospect of war deeply divided the country. In January 1917, President Woodrow Wilson told Congress that the United States must remain neutral—an idea that he had been promoting since the start of the war. In late January and February, German submarines began attacking and even sinking some American ships. Wilson went back to Congress to request a declaration of war. Still, many Americans remained torn.

Bernays talked his way into a position at the newly formed Committee on Public Information. It was an independent agency set up by the U.S. government to influence public opinion in support of the war. Bernays immediately went to work using his old tricks. He organized rallies at Carnegie Hall with freedom fighters from Poland, Czechoslovakia, and other countries; recruited Ford Motor Company and other American corporations to support the war and distribute pro-war brochures in their overseas offices; and planted U.S. propaganda in popular journals at home and abroad.

By the time the war was won, President Wilson clearly felt that Bernays had made a difference. He invited the twenty-six-year-old to accompany him to the Paris Peace Conference in January 1919. When Bernays arrived in Paris, he witnessed crowds flocking to President Wilson, a man he had positioned as a great liberator and defender of democratic freedom. "We'd worked to 'make the world safe for democracy,'" Bernays later said. "That was the big slogan."[15]

* * *

Bernays returned to the United States with a new awareness: "If you could use propaganda for war, you could certainly use it for peace."[16] In the following four decades, Bernays pulled off dozens of public relations coups.

When he was hired by a company that sold pork, he made bacon and eggs the all-American breakfast by getting his friend, a doctor, to write a letter to five thousand other doctors, urging them all to sign on to the recommendation that eating a heavier breakfast ("bacon and eggs") would be healthier for Americans.

He convinced kids to like taking baths by organizing soap-carving contests in public schools. That's because Procter & Gamble, his client, made a brand of soap (Ivory) that floated in water.

In the late 1940s he persuaded the U.S. government to build Route 66 as part of his work for Mack Trucks. More highway, more trucks.

Bernays seemed to understand that models influence desire. Doctors were the "expert" models who recommended bacon and eggs. Teachers modeled soap carving. And when Mack Trucks hired Bernays to defend the company against attacks from railroads, Bernays rallied legions of enthusiastic motorists, from the members of men's and women's driving clubs to milk delivery drivers and tire workers, to support the building of more highways.

But none of these things quite matched the magnitude of the coup that Bernays had pulled off decades earlier when he created one of the most powerful models of the century.

In 1929, the president of the American Tobacco Company, George Hill, approached Eddie Bernays with a tantalizing prospect: if he could break the taboo against women smoking in public, it could result in tens of millions of dollars in extra revenue per year. Hill was already paying Bernays a retainer of $25,000—a huge sum at the time (the equivalent of about $379,000 in today's dollars). If the campaign to get more women to smoke worked, some of the company's resulting profits would presumably go to Bernays.

Sales of the company's signature brand, Lucky Strike, were already exploding. During the war, soldiers' rations had included cigarettes. In the years after the war, smoking increased dramatically, as a generation of young men who first lit up as a comfort while facing the terrors of war were now hooked.

Women weren't part of the trend. There were societal taboos against them smoking in public, and men denigrated women who smoked even in private. The following quote from a male hotel manager in a 1919 *New York Times* article was typical:

> I hate to see women smoking. Apart from the moral reason, they really don't know how to smoke. One woman smoking one cigarette at a dinner table will stir up more smoke than a whole tableful of men smoking cigars. They don't seem to know what to do with the smoke. Neither do they know how to hold their cigarettes properly. They make a mess of the whole performance.

George Hill knew this taboo was hurting his profits. "If I can crack that market, I'll get more than my share of it," he told Bernays. "It will be like opening a new gold mine right in our front yard."[17] To do so, Bernays would have to effect a seismic shift in American culture strong enough to break sexist taboos.

Bernays first went to A. A. Brill, a disciple of his uncle Sigmund Freud and the leading psychoanalyst in the United States. Brill told Bernays that the cigarette is a phallic symbol that represents male sexual power. In order to turn cigarettes into an object important enough for women to fight for, Bernays would have to make smoking seem like a way for women to challenge male power. Cigarettes had to become, in Brill's words, "torches of freedom."

In order for that to happen, Bernays would have to give women a model.

Women's emancipation was in full swing in the 1920s. The Nineteenth Amendment, ratified in August 1920, had given women the right to vote; women were earning higher wages than ever before by taking jobs that had opened up during the war; and flappers were celebrating their new-found freedoms drinking French 75 cocktails and listening to Duke Ellington at the Cotton Club. The time was ripe for anything that appealed to freedom.[18]

Bernays concocted his plan. In March 1929, he identified the Easter

Day parade in New York City as the perfect opportunity to transform cigarettes into "torches of freedom." The parade was a spectacle of high fashion,
a media frenzy, an opportunity for well-to-do New Yorkers to strut down
Fifth Avenue to see and be seen.

An Easter procession down Fifth Avenue had emerged as early as the
1870s. Back then, Easter was as important to retail sales as Christmas is today. As the event became ritualized and turned into a parade, women wore
their best hats and colorful Easter dresses. As they exited the churches up
and down Fifth Avenue, they fell into step with other high-society women
using the street as their runway. They visited the flower-filled churches on
the parade route while being admired by the lower classes lined up along
the sides of the street to watch.

Bernays's plan was to convince a carefully selected group of these
women to light up Lucky Strike cigarettes defiantly during the parade,
the world's largest stage. It would be equivalent to a modern influencer
campaign of epic proportions: imagine if Beyoncé stopped her Super Bowl
halftime performance midsong, pulled out a Juul, and puffed away, with
the cameras zooming in on the brand and flavor.

According to Larry Tye's account, Bernays tapped a friend at *Vogue* who
helped him gather a list of thirty influential New York debutantes. He then
made an appeal to high-society women in New York City newspapers
through his friend Ruth Hale, a leading feminist.

A memo went out from his office with details of the event: "Women
smokers and their escorts will stroll from Forty-Eighth Street to Fifty-
Fourth Street on Fifth Avenue between Eleven-Thirty and One O'Clock."
Bernays planned to seed the event by hiring ten women to smoke conspicuously as they joined the parade, and he knew exactly what kind of woman
he wanted. "While they should be goodlooking, they should not look too
model-y," he wrote.

The memo continued:

Business must be worked out as if by a theatrical director, as for example:
one woman seeing another smoke, opens her purse, finds cigs but no
matches, asks the other for a light. At least some of the women should
have men with them.

Edith Lee smokes a cigarette
at the Easter Day Parade,
New York, 1929. (Photo courtesy
of the Library of Congress.)

At the appointed time, following Bernays's instructions, the models whipped out their Lucky Strikes and smoked them in their flapper hats and fur-trimmed coats as they walked flouncily down the street.

Bernays left nothing to chance. He made sure professional photographers and journalists were there to capture the moment. He even instructed them to use the phrase that he had selected, "torches of freedom," when describing it. He knew that the event would stir up debate, but how could anyone in this postwar world not be on the side of *freedom*?

As Tye tells it, pictures of women smoking their "torches of freedom" appeared on the front pages of major U.S. newspapers the next day: from the *New York Times* to the daily paper of Albuquerque, New Mexico.

United Press International mentioned a woman named Bertha Hunt, who "struck another blow on behalf of the liberty of women." She had pushed through the crowd in front of St. Patrick's Cathedral to lead the charge.

"I hope that we have started something," Hunt told journalists, "and that these torches of freedom, with no particular brand favored, will smash the discriminatory taboo on cigarettes for women and that our sex will go on breaking down all discriminations."[19] What she failed to mention while being interviewed was that she was Bernays's secretary and was reciting a carefully crafted statement.

Bernays orchestrated the entire Easter Day stunt to make it look like the women *spontaneously* started smoking, and that Hunt *spontaneously* desired to push through that crowd in front of the cathedral. He used the Romantic Lie against people.

He gave the illusion of autonomy—because that's how people think desire works. Models are most powerful when they are hidden. If you want to make someone passionate about something, they have to believe the desire is their own.

Within days, women took to the streets in cities across the United States lighting up their own "torches of freedom." Sales of Lucky Strikes tripled by the following Easter.

Mimetic Games

People play mimetic games because they have a tacit knowledge of how mimetic desire works even if they can't name it. Kids respect the laws of gravity long before they learn about them in school. Likewise, adults often play games of desire that exploit the vulnerabilities of others.

Let's take a brief look at how these games are played in romance, business, and advertising.

ROMANCE

When he was in his early twenties, during his university studies in France, René Girard got his first glimpse of mimetic desire in a surprising way: a roller coaster of desire with a girl he was dating. Girard recounted the incident to Robert Harrison during a 2005 episode of his Stanford radio show, *Entitled Opinions*.[20] The turning point in the relationship occurred when the woman suggested they get married. Right

Tactic 1

NAME YOUR MODELS

Naming anything—whether it's emotions, problems, or talents—gives us more control. The same is true for models.

Who are your models at work? At home? Who are the people influencing your buying decisions, your career path, your politics?

Some models are easy to name. They are what we typically think of as "role models"—people or groups we find exemplary, people we want to emulate in a positive way. We're not ashamed to acknowledge them.

Others we don't think of as models. Take fitness. A personal trainer is more than a coach—she is a model of desire. She wants something for you that you do not yet want for yourself enough to do what you need to do. The important shift is seeing people in more than their professional role but also in their role as an influencer of desire. This applies to your children's teachers, your colleagues, and your friends.

Harder to name are the people who come from inside our world and who might be modeling rivalrous or unhealthy behaviors—the people around whom we orbit without knowing it, who affect what we want. We'll explore this type of less-acknowledged model in Chapter 2 because they have come to dominate our world. Here's one way to identify them: think seriously about the people you least want to see succeed.

away, he experienced a decrease in desire and quickly backed off. They went their separate ways.

And then, as soon as he had moved away and she had accepted it— and, presumably, started entertaining a relationship with another man— Girard was drawn back to her. He pursued her once again, but she rebuffed him. The more she denied herself to him, the more he wanted her. "She influenced my desire by denying it," Girard said.[21]

It was as if *her lack of desire* for him affected the strength of *his desire* for her. What's more, the interest that other men showed in her affected him. They were modeling her desirability to him. Through her withdrawal from him, she was modeling it, too. "I suddenly realized that she was both object and mediator for me—some kind of model," Girard remembered.

People don't only model the desire for third parties or objects; they can also model the desire for *themselves*. Playing hard to get is a tried-and-true method to drive people crazy, but few ever ask why. Mimetic desire provides a clue. We are fascinated with models because they show us something worth wanting that is just beyond our reach—including their affection.

Fyodor Dostoevsky's book *The Eternal Husband* shows the comedy and tragedy of mimetic romance. A widowed man seeks out the former lover of his deceased wife—the man with whom she had an affair—because he subconsciously views this man as his romantic and sexual superior. The widowed man has found a new woman he would like to marry, but he can't do so without the assurance that his rival, his wife's former lover, also desires his new bride-to-be. He masochistically subjects himself to more humiliation at the hands of this other man, the Eternal Lover, while he remains the Eternal Husband. As long as he doesn't realize who is pulling the strings of his love life, he will continue to torture himself at the feet of his rival. He can't stop measuring himself against him.

The Eternal Husband is an extreme example of how mimesis can hijack relationships. It's usually not so obvious. Consider an insecure guy who feels sparks on his first few dates with someone. They both decide to get more serious. The first thing he does is introduce her to all his friends—because he desperately needs their approval. He's looking for some indication that at least one of them might want to be with her, too. When none of them seem interested, he begins to doubt that he made the right choice. He seeks validation for his choices from his models, like the Eternal Husband.

Or consider a sophomore in high school who posts a selfie to Instagram. She's beaming next to her new boyfriend at a sushi restaurant. Immediately, her ex—who broke up with her only a few weeks ago, confident in his decision, and whom she hasn't heard from since—starts texting her, confessing his love. "You don't know what you want," she tells him. "Make up your mind!" She's right: he didn't know what he wanted until he saw her with another guy—a senior, his older brother's age, who is going to the University of North Carolina on a basketball scholarship. Her renewed desirability has nothing to do with how she looks in her Instagram photo; it's a product of her being wanted by another man—and not just any man, but one who possesses all of the characteristics that her ex-boyfriend would like to have.

A friend and collaborator of Girard's, the psychoanalyst Jean-Michel Oughourlian, recommended a shocking tactic to people who came to him in his clinical practice complaining that their spouse no longer seemed interested in them: he would suggest they find someone to compete with the spouse for their time and attention. Even the remote suspicion that someone else might be competing for a spouse's time can be enough to arouse and intensify desire. (I'm not suggesting that anyone intentionally try to make their spouse jealous—although it seems to be a tactic that many people already use, quite naturally.)

Romance can feel like a roller-coaster ride because that's how mimetic desire moves.

RISKY BUSINESS

In his book *Give and Take: Why Helping Others Drives Our Success*, Wharton professor Adam Grant tells the story of veteran entrepreneur Danny Shader. He had already founded two successful companies and was looking to raise money for his next venture, his most exciting one yet.

At his daughter's soccer game in Silicon Valley, Shader runs into prominent venture capitalist (VC) David Hornik and tells him he's working on something new. They agree to meet. A few days later, Shader drives to Hornik's office and pitches him on the idea. Hornik sees the potential of the company right away. Within a week, he gives Shader a term sheet for an investment.

Unlike most VCs, though, Hornik doesn't give Shader an exploding offer—one that expires after a certain amount of time. VCs do this to apply the maximum amount of pressure on a founder to accept a deal. Founders, on the other hand, like to shop term sheets around to every VC possible, playing them against each other in the hopes of sparking a bidding war.

But Hornik is a different brand of VC. He puts no limits on his offer. He *encourages* Shader to talk to other VCs. He provides Shader with a long list of forty references attesting to his reputation as an investor. Hornik wanted Shader to choose him because he was the best partner and not because of the investment alone. So Shader made the rounds and talked to other VCs.

A few weeks later, he called Hornik and informed him that he had accepted funding from another investor. He told Hornik that he had been so pleasant, so nonthreatening and affable, that he worried about whether Hornik would challenge him in the boardroom and push him to be his best. "My heart said to go with you," Shader said, "but my head said to go with them."[22]

Shader was caught up in a mimetic value game. The investors who modeled their own desirability—who postured as selective and demanding—took on a higher value in his mind than the one investor who didn't. Elite colleges don't keep their admissions rates low because they have to; they keep them low to protect the value of their brands.

Hornik didn't play the games that other VCs liked to play. He wanted to work with founders who recognized the value of what he truly had to offer, not the value assigned to him through the charade of a VC beauty contest.

The story of Shader and Hornik serves as a warning to stand on guard against mimetic valuations. It's the Paradox of Importance: sometimes the most important things in our lives come easily—they seem like gifts—while many of the least important things are the ones that, in the end, we worked the hardest for.

ADVERTISING IRONY

Today's manipulators of desire aren't as brash as Eddie Bernays. They've wised up. To some extent, so have we.

Advertising gurus know that we scrunch our noses when we're being sold something too hard. They know that they can no longer capture us by simply showing us a beautiful and happy-looking person drinking a particular brand of soda. For the past thirty to forty years at least, advertisers have had to use a different and less on-the-nose tactic: irony. They make fun of themselves to lower our defenses.

A 1985 Pepsi commercial portrays a man pulling up to a beach in a van and broadcasting his lusty gulps of ice-cold Pepsi from speakers mounted on the roof until everybody on the beach follows one another to the van, where the guy proceeds to sell them all bottles of Pepsi. The commercial closes with the words: "Pepsi: The Choice of a New Generation." The use of the word "choice" is ironic because the commercial portrays all the people schvitzing on the beach as having little to no choice at all.

The goal is getting people to think, "Oh, those lemming-like, silly people in the commercial." The moment a person exempts themselves in their own mind from the very thing they see all around them is the moment when they are most vulnerable. As David Foster Wallace pointed out, "Joe Briefcase," sitting on his couch watching the Pepsi commercial alone, thinks he has transcended the mass of plebeians that Pepsi must be advertising to—and then he goes out and buys more Pepsi, for *reasons that he thinks are different.*[23]

And if he doesn't drink more Pepsi, then he will be more likely to drink something else that he feels separates himself from the masses—maybe kombucha. The consumption can also be of something besides a soft drink, something quite different in type: the latest Netflix Original documentary, say, or podcasts that make him feel smarter than his friends. The pride that makes a person believe they are unaffected by or inoculated against biases, weaknesses, or mimesis blinds them to their complicity in the game.

If a news organization can convince its viewers that its programming is neutral, it disables their defense mechanisms. Big Tech companies do something similar. They present their technology as agnostic—as just a "platform." And that's true, so long as we evaluate it in a materialistic way, as bits and bytes. Yet, on a human level, social media companies have built engines of desire.

Mimetic models lie in wait every time we glance at our phone. The families of childhood friends post photos in which every day looks like a Christmas card, and Instagram models with bleached white teeth show us how they eat their nutritious breakfasts. The universe of desire is dotted with billions of stars who appear to shine brightest at the exact moment when we find it hardest to see.

Models That Move Markets

On February 3 and 4, 2020, shares of stock in the electric automaker Tesla had a parabolic rise of more than 50 percent. This capped a four-month period in which the stock had already doubled in price. By the end of the day on February 4, it had nearly quadrupled over this four-month period.

Tesla had been a publicly traded company for nearly a decade. It was not a newly listed IPO, where that kind of price action and volatility would be less of an anomaly. What news was moving the stock?

Nothing extraordinary. The company beat third- and fourth-quarter earnings estimates. There was some good news about its China factory. But it was hard to see what was driving the wild movements.

Before October 2019, Tesla had seemed like it might be on the brink of disaster. CEO Elon Musk had roiled markets with his erratic behavior, and investors were wondering how the company could stay afloat while burning through almost $3.5 billion in cash in twelve months. Tesla reported losses of over $1 billion in the first half of 2019. And then came this out-of-this-world explosion in the company's stock price.

Professional analysts were baffled. "I talked to people at Goldman Sachs, who are usually the world's greatest experts on explaining stock prices, and they're now asking me whether I have any idea what the heck is going on with Tesla stock," said Bob Lutz, a former auto industry executive, on a BBC *Business Daily* podcast. No one believed the stock's movement corresponded to "reality."[24]

But it did—just not the reality that most analysts accept. To paraphrase Shakespeare: there are more things in heaven and earth than are dreamt of in their investment philosophies.

One version of the Romantic Lie in finance is the "efficient market" hypothesis (closely related to "rational expectations," another hypothesis). Efficient market theory is the belief that asset prices are functions of all available information. Company news, investor expectations, current events, political news, and everything else that might affect a company's valuation are all assumed to be perfectly reflected in the stock price. The price changes as a function of time, as new information becomes available.

But there is more to understanding markets, and people, than information.[25]

A couple of pieces of data should have alerted investors in Tesla stock that more than information was driving the stock price. On February 4, the second day of the rally, more than $55 billion of Tesla stock changed hands—more than any stock in history at the time. On the same day, people who started searching Google with the words "Should I" received an auto-suggested completion of their question: "Should I buy Tesla stock?"

Millions of people were searching Google to find out whether they should buy Tesla based on whether other people wanted to buy Tesla. This, in my view, is not *merely* information. It's mimetic desire.

Desire is not a function of data. It's a function of other people's desires. What stock market analysts referred to as "mass psychosis" was not so psychotic after all. It was the phenomenon of mimetic desire that Girard had discovered more than fifty years earlier.

In both bubbles and crashes, models are multiplied. Desire spreads at a speed so great we can't wrap out rational brains around it. We might consider taking a different, more human, perspective.

"Conformity is a powerful force that can counteract gravity for longer than skeptics expect," writes *Wall Street Journal* finance columnist Jason Zweig. "Bubbles are neither rational nor irrational; they are profoundly human, and they will always be with us."[26]

Mimetic desire is profoundly human, and it will always be with us. It's not something "out there" to be engineered and life-hacked away. It lives loudly within us, closer than we can see with our own eyes.

And the children go to summer camp
And then to the university
Where they are put in boxes
And they come out all the same.

—Malvina Reynolds

Chapter 2

DISTORTED REALITY
We're All Freshmen Again

Celebristan . . . Pseudonyms . . . Cat Worship . . . Signified
Rappers . . . Mirrored Imitation

> Here we go again. Perception trumping reality once more.
> —Dick Fuld, former CEO of Lehman Brothers, while watching
> breaking news about the firm's impending demise[1]

During the 1972–1973 school year, three years before the first personal computer was introduced, a freshman at Reed College in Oregon was looking to make some cash. He learned a lesson that would help him one day become the greatest showman of his generation.

Steve Jobs had arranged to sell his old IBM Selectric typewriter to a fellow student named Robert Friedland. Like Jobs, Friedland was an undergraduate—but he was four years older than Jobs. He had been kicked out of Bowdoin College in Maine and sentenced to two years in prison for possessing $125,000 worth of LSD. After he made parole, he enrolled at Reed and made plans to run for student body president and travel to India to meet a Hindu guru. But first he needed a typewriter.

Steve Jobs didn't know anything about his buyer. He arrived at Friedland's room to deliver the goods and collect his cash, but nobody responded to his knocks on the door. He tried the handle. It was unlocked. To save the trip, he considered leaving the typewriter inside and collecting the money later. So he opened the door.

When he walked in, he was mortified by what he saw: Friedland was on the bed having sex with his girlfriend. Jobs tried to leave, but the stranger invited him to sit down until he was done. *This is kind of far-out*, Jobs thought.[2]

Who was this creature who appeared to lack any sense of taboo, any sense that his behavior would make normal people cringe? Who seemed to do exactly what he wanted and make no apologies?

Fellow student Daniel Kottke—who went on to become one of the first employees at Apple—would later remark on the influence Friedland had on Jobs. Friedland "was mercurial, sure of himself, a little dictatorial," he remembered, according to Walter Isaacson's biography of Jobs. "Steve admired that, and he became more like that after spending time with Robert."

By the time he started Apple, Jobs had become known for his own quirky behavior. He walked around the office barefoot, rarely took showers, and enjoyed soaking his feet in the toilet.

"When I first met Steve he was shy and self-effacing, a very private guy," says Kottke. "I think Robert taught him a lot about selling, about coming out of his shell, of opening up and taking charge of a situation."

Jobs had not realized it, but at the moment he walked into that room in college, Friedland had become a model to him. Jobs would later come to see through Friedland, but Friedland's immediate impact on the young Jobs was formative. He taught Jobs that strange or shocking behavior mesmerizes people. People are drawn to others who seem to play by different rules. (Reality TV exploits this.)[3]

As Jobs became a skilled practitioner of this behavior, his colleagues described him as having a "reality distortion field." Jobs seemed to be able to bend everyone in his orbit to his will—that is, to his desires. The reality distortion field extended to anyone in close proximity to him. To what can we attribute his effect on people?

He was brilliant, but that's not what made him so alluring. The Enlightenment philosopher Immanuel Kant was brilliant, too, but his life was so mundane that townspeople could set their watches to his daily walks. Jobs was mesmerizing because he *wanted* differently.

We often attribute a person's magnetism to some objective quality about them—a manner of speaking, intelligence, tenacity, wit, or confidence. Those things help, but there is more.

We are generally fascinated with people who have a different relationship to desire, real or perceived. When people don't seem to care what other people want or don't want the same things, they seem otherworldly. They appear less affected by mimesis—anti-mimetic, even. And that's fascinating, because most of us aren't.

Two Kinds of Models

Nobody likes to think of themselves as imitative. We value originality and innovation. We are attracted to renegades. But everybody has hidden models—even Steve Jobs.

In this chapter, we'll explore two kinds of models that affect us in different ways: those who are *outside* of our immediate world and those who are *inside* of it. Mimesis has different consequences in each case. Which world do you think Robert Friedland was in? By the end of this chapter, you'll see that the answer is not so simple.

There's something strange about our relationship to imitation. Aristotle recognized nearly 2,500 years ago that humans possess advanced capabilities of imitation that allow us to create new things. Our ability to imitate in complex ways is why we have language, recipes, and music.[4]

Isn't it odd, then, that people generally frown upon imitation? One of our greatest human strengths is seen as a source of embarrassment, a sign of weakness, even something that can get you into trouble.

Nobody wants to be known as an imitator—except in very specific cases. We encourage children to imitate role models, and most artists generally recognize the value of imitating the masters. But imitation is totally taboo in other circumstances. Imagine if two friends started showing up at every social gathering wearing matching clothes; if a person who received a gift always reciprocated by giving the other person the same gift they were given; if someone constantly mimicked the accent or mannerisms of coworkers. These things would be considered strange, rude, or insulting, if not infuriating.

There's something a little uncomfortable about a friend who gets a haircut too much like our own.

To make matters even more confusing, why is it that in most organizations imitation seems to be simultaneously encouraged and discouraged? Dress like someone in a position that you'd like to be in, but not too closely;

imitate the cultural norms, but make sure you stand out; emulate key leaders in the organization, but don't seem like you're brown-nosing.

Elizabeth Holmes, the former CEO of the now-defunct biotech company Theranos, openly imitated Steve Jobs. She wore black turtlenecks and hired every Apple designer she could find. But imagine if a junior employee at Theranos started mimicking Holmes, walking around in black turtlenecks, sporting blue contact lenses, mimicking Holmes's intense stare, even speaking in Holmes's low pitch and dry style. What do you think would happen? They'd lose their job.

It's as if everyone is saying, "Imitate me—*but not too much*," because while everyone's flattered by imitation, being copied too closely feels threatening.

In this chapter, you'll see why.

This is the most technical chapter in the book. It's where we'll lay the groundwork for understanding the most important implications of mimetic theory.

First we'll see how desire is affected differently by people who are at a great social distance from us (celebrities, fictional characters, historical figures, maybe even our boss) and those who are close (colleagues, friends, social media connections, neighbors, or people we meet at parties).

In the first situation, where there is a large difference in status, models live in a place we'll call Celebristan. From where I stand, residents of Celebristan include Brad Pitt, LeBron James, Kim Kardashian, and the founders of unicorns (start-ups with billion-dollar valuations). These people might as well live in a different universe of desire; there is little chance of their desires coming into contact with mine. There is some social or existential barrier that separates us.

There isn't always such an obvious and dramatic difference. A managing director at an investment bank appears to live in Celebristan to an analyst; a priest to a layman; a rock star to the backup singers; Tony Robbins to his seminar attendees; even an older brother to a younger brother in many cases. Celebristan is where models live who mediate—or bring about changes in our desires—from somewhere *outside* our social sphere, and with whom we have no immediate and direct possibility of competing on the same basis.

We're more threatened by people who want the same things as us than by those who don't. Ask yourself, honestly: whom are you more jealous of? Jeff Bezos, the richest man in the world? Or someone in your field, maybe even in your office, who is as competent as you are and works the same amount of hours you do but who has a better title and makes an extra $10,000 per year? It's probably the second person.

That's because rivalry is a function of proximity. When people are separated from us by enough time, space, money, or status, there is no way to compete seriously with them for the same opportunities. We don't view models in Celebristan as threatening because they probably don't care enough about us to adopt our desires as their own.

There is another world, though, where most of us live the majority of our lives. We'll call it Freshmanistan. People are in close contact and unspoken rivalry is common. Tiny differences are amplified. Models who live in Freshmanistan occupy the same social space as their imitators.

We're easily affected by what other people in Freshmanistan say or do or desire. It's like being in our freshman year of high school, having to jostle for position and differentiate ourselves from a bunch of other people who are in the same situation. Competition is not only possible, it is the norm. And the similarity between the people competing makes the competition peculiar.

This chapter is about why imitation differs in kind and in quality depending on where it happens. I will give you a tool kit for understanding how people come under the influence of certain models, how models distort reality, and why mimetic desire is so dangerous in Freshmanistan.

CELEBRISTAN

René Girard calls models in Celebristan *external mediators of desire*. They influence desire from outside of a person's immediate world. From the perspective of their imitators, these models possess a special quality of being.

Dream dates live in Celebristan so long as you've never even met the person you dream of dating, or if the dream date lives in an untouchable social sphere that puts them out of reach. Celebrities who agree to attend a high school prom are sweet—but everyone understands that the high school kid isn't going to take them away from their A-list suitors. The phrase "out of your league" hints at this strange, unattainable world.

In Celebristan, there is always a barrier that separates the models from their imitators.[5] They might be separated from us by time (because dead), space (because they live in a different country or aren't on social media), or social status (a billionaire, rock star, or member of a privileged class).

Julia Child is a model to millions who aspire to elevate their home cooking, and Abraham Lincoln is a model to many politicians. Since both are dead, they occupy a permanent place in Celebristan. There is no chance that they will enter into our world and become our rivals.

This brings us to an important feature of Celebristan models: because there's no threat of conflict, *they are generally imitated freely and openly.*

In 1206, Francesco Bernardone, a twenty-four-year-old man from a wealthy merchant family, stripped naked in the central piazza of his town in central Italy, handed his fine clothes to his father, and renounced his hereditary rights. For the next eight hundred years, hundreds of thousands of people followed the path of Saint Francis of Assisi. They took radical vows of poverty, imitated his devotions, and wore the same style of plain brown robe. At the time of this writing, there were around thirty thousand Franciscans in the world. And in 2013, a former cardinal of Buenos Aires, Jorge

CELEBRISTAN

Mario Bergoglio, became the 266th pope and declared that his papal name would be Francesco—to signal that he intended to imitate Saint Francis of Assisi's focus on the poor.

Saints become Celebristan models—declared worthy of imitation—only after they are dead. Nobody can officially be declared a saint while they are alive. A similar protocol is in place with a Hall of Famer in professional sports—no athlete can be inducted into the Hall of Fame while still playing the game. People only truly become legends after they retire or die because they enter into a different existential space.

Some models use a trick to cement their Celebristan citizenship: they guard their identities to heighten our sense of intrigue. Banksy, J. D. Salinger, Stanley Kubrick, Elena Ferrante, Terrence Malick, and Daft Punk all have hidden themselves from view, which makes them appear to exist in a different plane.

Satoshi Nakamoto, the pseudonymous programmer thought to be the inventor of Bitcoin, boosted his mimetic value into the upper stratosphere of Celebristan through secrecy. He made himself impossible to compete with. "You can't be Satoshi-but-more-charismatic, because nobody knows for sure that they've met him," write Tobias Huber and Byrne Hobart. "You can't be Satoshi-but-more-paranoid, because he still hasn't been firmly identified. You can't be Satoshi-but-more-forward-thinking, unless you start now and develop something that's a bigger deal than Bitcoin ten years from now, without getting caught."[6]

Hierarchies in companies can create barriers to competition, making it practically impossible for some people to compete with others for the same roles and accolades. From the perspective of a call center employee working for a hierarchical company, a C-suite executive might as well live on a different planet. The CEO is rarely seen. She's untouchable. There's no serious threat of a customer service rep competing with her for her job in the near term.

The same is true of many founders and employees, teachers and students, professionals and amateurs in a sport. (The distinction between amateurs and professionals is marked by a rite of passage, which exists to demarcate who can compete with whom.) In Celebristan, people don't compete with their imitators. They may not even know they exist. This makes it a relatively peaceful place.

In Freshmanistan, however, fierce competition can arise between any two people at any time.

FRESHMANISTAN

Freshmanistan is the world of models who mediate desire from *inside* our world, which is why Girard calls them *internal mediators of desire*. There are no barriers preventing people from competing directly with one another for the same things.

Between social media, globalization, and the toppling of old institutions, most of us are living nearly our entire lives in Freshmanistan.

Friends live in Freshmanistan together. Shakespeare's play *The Two Gentlemen of Verona* shows how easily desires in this world become intertwined. Valentine and Proteus, friends since childhood, discover their desires converging on the same woman—not incidentally, but *because* of the other one's desire. Proteus is in love with a girl named Julia. When he goes to visit his friend Valentine in Milan, Valentine talks up his own new love interest, Silvia. Upon hearing his friend's excessive praise, Proteus instantly falls in love with Silvia, too. A day earlier Proteus had pledged his eternal love to Julia; now he wants Silvia. Shakespeare often portrayed mimetic desire in comedies because it's more palatable for people that way—they can laugh at others behaving ridiculously from a safe distance, without being reminded of their own mimesis.

Mimetic desire is both the bond and the bane of many friendships. A common example: one friend introduces the other to baking; the desire to become a better baker is then shared by both friends, which leads

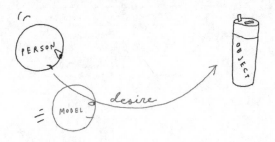

FRESHMANISTAN

them to spend more time together baking. But if the friendship becomes tinged with mimetic rivalry, it can lock them into a never-ending game of rivalrous tug-of-war that extends beyond baking to relationships, career success, fitness, and more. The same force that drew them together, mimetic desire, now pushes them apart as they try to differentiate themselves.

Remember what it's like to be a freshman in high school? People from many different backgrounds are thrust into the same building, same hallways, same classrooms. Skaters get assigned to projects with thespians; rockers play sports with jocks; jocks sit next to nerds.

It might seem like these groups are very different. Don't nerds look at jocks as if they exist in another world? Yes. But they are far more alike than different. They're roughly the same age; they're all dealing with adolescent hormones. They attend the same classes and eat in the same lunchroom. Any one of them can come into contact with any other on any given day.

Everyone is taking mimetic cues from everyone else, but almost nobody knows it. An unspoken battle of differentiation occurs as each person tries to carve out an identity over and against the rest.

Celebristan (World of External Mediation)	Freshmanistan (World of Internal Mediation)
Models are distant in time, space, or social status	Models are close in time, space, or social status
Difference	Sameness
Models are easy to identify	Models are hard to identify
Open imitation	Secret imitation
Models acknowledged	Models unrecognized
Relatively stable, fixed models	Unstable, constantly changing models
No possibility of conflict between models and imitators	Conflict between models and imitators is normal
Positive mimesis* is possible	Negative mimesis is the norm

* Positive mimesis is explained in Part II of this book.

Distortions of Reality

Being a literal freshman is disorienting and filled with anxiety. So is life in Freshmanistan. Reality is distorted in many ways. Here are a few of them.

DISTORTION 1: THE MISAPPROPRIATION OF WONDER

People exaggerate the qualities of their models constantly, whether the models are in Freshmanistan or Celebristan. When a model is from Celebristan, people openly gawk at and admire them. An obvious case of this is people mobbing a celebrity for a signature. A less obvious case is the way that a professor might openly admire another professor who chairs a department at a different university. But the moment the second professor transfers to the first professor's own department, the dynamic changes immediately. Now they occupy the same world. Now they have to compete for the same things.

In Freshmanistan, people occupy the same space as their models, so they have to wonder at them in secret amazement. They could never admit the embarrassing truth that they want to be more like their neighbors, colleagues, or maybe even one of their friends. In Freshmanistan, there is an omertà. Steve Jobs fell under the spell of Robert Friedland because he wanted, in some way, to be more like him. The fellow student had already colonized his world of desire.

Girard calls this striving not for any particular *thing* but for some new way of living or being *metaphysical desire*.[7] In Greek, the word *meta* means "after." Aristotle had studied the physical world and learned all that he could learn about it. Then he asked, "What now?" He applied himself to the study of what would later be called *metaphysics*, which literally means "after the physical."[8]

Girard believed that all true desire—the post-instinctual kind—is metaphysical. People are always in search of something that goes beyond the material world. If someone falls under the influence of a model who mediates the desire for a handbag, it's not the handbag they are after. It's the imagined newness of being they think it will bring. "Desire is not of this world," Girard has said, ". . . it is in order to penetrate into another world that one desires, it is in order to be initiated into a radically foreign existence."[9]

The metaphysical nature of desire leads to strange distortions in the way that we see other people. Girard sees this happening in the tragic case of anorexia nervosa and bulimia. The desire to be like a model who represents an ideal body image is stronger than the need for basic sustenance. These are obviously psychological diseases, but Girard doesn't think we've properly accounted for the role of mimetic desire in their etiology. In his view, they are cases of metaphysical desire overpowering physical needs.[10]

We all suffer from this problem in our own way—we are all, in some way, anorexic, looking for models who can satisfy a nonphysical hunger, a metaphysical desire.

THE WORSHIP OF CATS

We are most likely to take someone as a model if they don't seem to be suffering from desire like us. Consider the cat. Where does its allure come from? Why did the Egyptians worship cats?

The reasons are complex. But mimetic theory offers a hint: cats seem far less needy than we are. They don't want the way we do. The Egyptians may have associated the attitude of cats with various divinities because of the cat's apparent lack of desire. Who is less needy than a god?

Sure, some cats meow as if they're starving until you feed them. Some cats can't get enough snuggles. But cats are mercurial. They usually seem uninterested in your opinion—much the way Steve Jobs didn't care how anyone else felt about his not showering and putting his feet in the toilet.

When I used to scold my German shepherd puppy for tearing apart my couch, he lowered his eyes and slinked away. By contrast, if I yell at my cat for tearing up my couch when I'm gone, she sticks her butt in my face and struts out of the room. When I try my best to get her to come, she sits down and licks her paw. The effect is similar to a person who pretends to be uninterested in anyone's attention and approval. The affectation of self-sufficiency makes them fascinating.

While we toil away to fulfill our ever-changing desires, the cat grooms itself. It needs nothing, wants nothing.

So don't always try to pet a cat when you encounter one on the street, as popular psychologist Jordan Peterson advises in his *12 Rules for Life*. Rather, make the cat *want* to get petted when it encounters you. Then you'll have really accomplished something special.

DISTORTION 2: THE CULT OF EXPERTS

One hundred years ago, there was a much wider gap in knowledge between someone who had a doctoral degree and someone who didn't. Today, with the world's information at nearly everyone's fingertips, the knowledge gap between people with a great amount of formal education and those with less has narrowed. In fact, holding certain degrees, such as a PhD or an MBA, can count against you if you're pursuing jobs at companies that view them as a sign of one's complacency. We are witnessing an inversion of value.

Peter Thiel instituted the Thiel Fellowship in 2011 to pay promising entrepreneurs to start businesses instead of going to college. The fellowship was able to make its value proposition attractive in part because it intentionally hacked mimetic desire: *getting a fellowship was harder than getting into Harvard.* (The first class of fellows had an acceptance rate of around 4 percent, and in subsequent years it went down to around 1 percent.) Among the dropouts funded by the fellowship were brilliant, ambitious kids like Vitalik Buterin, co-creator of the decentralized open-source blockchain Ethereum, and Eden Full, inventor of a technology that enables solar panels to follow the sun. These entrepreneurs modeled something even more important than a Harvard degree to many young people; they modeled a different track.

Today value is largely mimetically driven rather than attached to fixed, stable points (like college degrees). This has created opportunities for anyone who can stand out from the crowd. This has positive and negative consequences.

People are desperate to find something solid to hold on to in today's "liquid modernity" (to borrow a term from sociologist and philosopher Zygmunt Bauman). Liquid modernity is a chaotic phase of history in which there are no culturally agreed-upon models to follow, no fixed points of reference. They have melted like glaciers and plunged us into a stormy sea with limited visibility. Celebristan is collapsing into it.

At the same time, the world is becoming increasingly complex. Think of the global financial system. The proportion of total available knowledge that any single person has is microscopic. So we rely more than ever on models, like hedge fund manager Ray Dalio, to make sense of it. Radical individualism does not free people from needing models. But where will they come from?

"Since modern man has no way of knowing what is going on beyond himself, since he cannot know everything, he would become lost in a world as vast and technically complex as ours, if he had really no one to guide him," wrote Girard in his book *Resurrection from the Underground: Feodor Dostoevsky*. "He no longer relies on priests and philosophers, of course, but he must rely on people nevertheless, more than ever, as a matter of fact."

And who are these people? "They are the *experts*," continues Girard, "the people more competent than we are in innumerable fields of endeavor."

Experts are the ones who help mediate desires, who tell us what is worth wanting and what is not. Tim Ferriss shows millions of people which books to read, which movies to watch, and which new apps to use in his "5-Bullet Friday" emails (which I never miss). He's an expert. He even teaches others how to hack expertise. "Expert status can be created in less than four weeks if you understand basic credibility indicators," he writes.

Katie Parla is the expert source for people who want to learn about restaurants in Rome. Marie Kondo is the expert purger of junk. Richard Blevins—better known as "Ninja"—expertly plays video games with over 650,000 people watching at one time.[11]

Much as Paris Hilton and the Kardashians became known as "being famous for being famous," there are now experts who fill booking slots on cable news programs who are experts at being experts. Actor and writer Dax Shepard even seems to be poking fun at the notion with his popular podcast, *Armchair Expert*, in which he interviews a wide-ranging assortment of guests from the standpoint of one who has no subject-matter expertise. At the end of every episode, his co-host, Monica Padman, does a "fact check" of the claims Shepard made during the interview. Only experts, of course, have the facts.

"The modern world is one of experts," wrote Girard. "They alone know what is to be done. Everything boils down to choosing the right expert."[12] If my friend is more on top of global affairs, urbanism, culture, and design than I am, it's because he has a *Monocle* subscription. If someone has better insights into the role technology is playing in our lives, it's because they listen to the right podcast (*Note to Self* by Manoush Zomorodi, by the way). And how's your cooking? I recommend Samin Nosrat.

There's such demand for new models that we insert mediators of desire where they don't belong—like *Shark Tank*–style business pitch competitions where experts, rather than the market, decide whether a business is valuable. We're model addicts. Right now, the models we prefer are experts.

That could be because we think of ourselves as more rational than ever—and we are, in many ways. Scientific progress has been swift in the past hundred years. However, we underestimate the strong role that mimesis plays in the way we choose our experts.

What is our basis for taking a source as authoritative? Is it because we checked all of the person's credentials? Is it because the source was fact-checked by Peter Canby's team at the *New Yorker*? Or is it because the person has the most followers on social media and a "Verified" sticker next to their name? Authority is more mimetic than we like to believe. The fastest way to become an expert is to convince a few of the right people to call you an expert.

The cult of saints has become the cult of experts. That doesn't mean we no longer rely on models to figure out what to want. It means that in a post-Enlightenment world, the preferred models are often those who seem most enlightened: the experts.

Models promise a kind of secret, salvific wisdom reminiscent of the early religious sect of gnosticism, which held the belief that one could be saved from predominating ignorance through an evolution in consciousness provided by "Messengers of Light." (Do you drink regular coffee? Then you obviously haven't read Dave Asprey, who knows that the beans you drink are covered in molds that produce mycotoxins, and you should buy his Bulletproof coffee to save yourself from the fate of plebeian coffee drinkers who are ignorant of this.) There's a model for everyone—someone ready to impart the specific knowledge needed to be happy and make people feel that they have escaped the fate of the masses. But any model who bills themself as this kind of expert is a charlatan.

Every once in a while, then, it's good to deconstruct the mimetic layers behind someone's authority and think seriously about how we chose our sources of knowledge in the first place. We might find that the road to our favorite experts was paved with mimetic influence.

Tactic 2

FIND SOURCES OF WISDOM THAT WITHSTAND MIMESIS

Experts play an increasingly prominent role in our society. But what makes an expert? A degree? A podcast? Increasingly, experts are crowned mimetically, like fashion.

Because there is less and less agreement about cultural values and even about the value of science itself (consider the debates about climate change), people find "experts" whose expertise is largely a product of mimetic validation. It's critical to cut through mimesis and find sources of knowledge that are less subject to mimesis.

Find sources that have stood the test of time. Be wary of self-proclaimed and crowd-proclaimed experts.

It's less likely that experts will be mimetically chosen in the hard sciences (physics, math, chemistry) because people have to show their work. But it's easy for someone to become an overnight expert on "productivity" merely because they got published in the right place. Scientism fools people because it is a mimetic game dressed up as science.

The key is carefully curating our sources of knowledge so that we are able to get down to what is true regardless of how many other people want to believe it. And that means doing the work.

DISTORTION 3: REFLEXIVITY

The billionaire investor and activist George Soros has claimed that financial markets operate according to the *principle of reflexivity*. "In situations that have thinking participants, there is a two-way interaction between the participant's thinking and the situation in which they operate," he writes in his book *The Alchemy of Finance*. Reflexivity in markets is partly what leads to market crashes and bubbles. Investors perceive there might be a crash, so they behave in a way that precipitates the crash.

Soros is rumored to have made over $1 billion in a single day by understanding this principle. In 1992, as the British government was spending large amounts of money to prop up the British pound, Soros bet up to $10 billion that they wouldn't be able to keep it stable. Soros's bet signaled to other investors that the smart money was against the British government, which in turn caused additional downward pressure on the currency. Eventually, the government caved and allowed the currency to float freely. It depreciated 25 percent against the U.S. dollar in a single day, yielding Soros a huge profit.

While Soros focuses on the principle of reflexivity in financial markets, it operates in many other domains of life. People worry about what other people will think before they say something—which affects what they say. In other words, our perception of reality *changes reality* by altering the way we might otherwise act. This leads to a self-fulfilling circularity.

This principle affects public and personal discourse. The German political scientist Elisabeth Noelle-Neumann coined the term "spiral of silence" in 1974 to refer to a phenomenon that we see often today: people's willingness to speak freely depends upon their unconscious perceptions of how popular their opinions are. People who believe their opinions are not shared by anyone else are more likely to remain quiet; their silence itself increases the impression that no one else thinks as they do; this increases their feelings of isolation and artificially inflates the confidence of those with the majority opinion.

Even clothes are reflexive, according to the author Virginia Woolf: "Vain trifles as they seem, clothes have, they say, more important offices than merely to keep us warm. They change our view of the world and the world's view of us. . . . There is much to support the view that it is clothes that wear us and not we them; we may make them take the mould of arm or breast, but they mould our hearts, our brains, our tongues to their liking."[13] Winston Churchill spoke about the reflexivity of architecture when he said, "We shape our buildings; thereafter they shape us."[14]

The principle of reflexivity has been unexplored in the domain of desire. We might reformulate Soros's definition of reflexivity like this: *In situations where desirous participants have the possibility of interacting with each other, there is a two-way interaction between the participants' desires.*

The situation is like being on a trampoline with another person jumping right next to you: neither person can jump without affecting the other. The reflexivity of desire in Freshmanistan distorts reality because people think they want things for spontaneous, rational reasons—the Romantic Lie—even while they are being affected by the people around them. This makes things seem different than they really are.

From 2003 to 2016 investors gave the previously mentioned Elizabeth Holmes, who liked to imitate Steve Jobs, more than $700 million. Her

MIMETIC RIVALRY

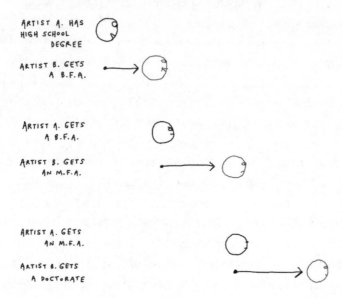

ARTIST A. HAS
HIGH SCHOOL
        DEGREE

ARTIST B. GETS
A B.F.A.

ARTIST A. GETS
A B.F.A.

ARTIST B. GETS
AN M.F.A.

ARTIST A. GETS
AN M.F.A.

ARTIST B. GETS
A DOCTORATE

company, Theranos, peaked at a valuation of over $10 billion. The investor funds allowed her to build a sleek Silicon Valley headquarters, hire coveted former Apple employees, and fuel a public relations campaign that landed her a lucrative contract with Walgreens. All this made new investors froth at the mouth to get in on the action. This kind of funding process is doubly mimetic: new investors want in because other smart investors are already in, and investor demand for the company's shares allows the company to tell a better story, which fuels even more investor demand.

The reflexivity of desire is most apparent in *rivalrous* relationships. When a person is focused on what a rival model wants, the desires of both individuals are reflexive. Neither can want anything without affecting the other's desire for it.

In Freshmanistan, a mimetic rivalry is like two people trying to race each other inside of the same car: Nobody gets ahead, and eventually they crash.

SIGNIFIED RAPPERS

The rivalry between the East Coast and West Coast in 1990s American hip-hop is a case study in the reflexivity of rivalrous mimetic desire.

In 1991, a virtually unheard-of rapper in the Bronx named Tim Dog released an album on which he adopted an angry and combative rap style that directly attacked West Coast rappers including Eazy-E, Dr. Dre, DJ Quik, and Ice Cube. Tim Dog was angry that West Coast record labels seemed to be ignoring the East Coast and disrespecting the quality of its hip-hop music. With a single song, he trolled West Coast rappers and sucked them into a mimetic rivalry.

At the end of 1992, West Coast rapper Dr. Dre dropped his debut album *The Chronic*, which went on to become one of the best-selling rap albums of all time. On it, the up-and-coming West Coast rapper Snoop Dogg mentions Tim Dog by name as retribution. I'll spare you the details of Snoop's verbiage here. Suffice it to say that he escalated the conflict.

The East Coast responded in kind. In 1993, Sean Combs signed the Notorious B.I.G. (known more widely as Biggie Smalls) to his upstart record label, Bad Boy Records. Biggie's song "Who Shot Ya?," released on the B-side of a single, was interpreted by young West Coast rapper Tupac Shakur as mocking him. Tupac had recently been the victim of a gunpoint robbery during which he was shot. Shortly after, he was signed to the controversial music label Death Row Records.

A war of escalating conflict ensued. For a few years in the mid-1990s, it seemed like every major song put out by the Bad Boy and Death Row record labels was in response to a song put out by the other. The mimetic rivalry between Tupac and Biggie ended with both of them dead.

When mimesis is strong enough, rivals forget about whatever objects they were fighting for in the first place. Objects become completely interchangeable—the rivals will fight for anything, so long as their opponent wants it. They become locked in a double bind—each reflexively bound to the desires of the other, unable to escape.

MIRRORED IMITATION

Why do all hipsters look alike, and why does nobody identify themselves as one?

The answer is *mirrored imitation*. Mirrors distort reality. They flip the sides on which things appear: your right hand appears on the left side in the mirror, and your left hand appears as if it's on the right side. The mirror image is, in some sense, an image of opposites. Mirrored imitation, then, is imitation that does the opposite of whatever a rival does. It is reflexive to a rival by doing something different from what the rival models.

When mimetic rivals are caught in a double bind, obsessed with each other, they go to any length to differentiate themselves. Their rival is a model for what *not* to desire. For a hipster, the rival is popular culture—he eschews anything popular and embraces what he believes to be eclectic, but he does so according to new models. According to Girard, "the effort to leave the beaten paths forces everyone into the same ditch."[15]

Reflexive, mirrored imitation is funny to those watching. No television show represents mimetic desire better than *Seinfeld*. In the episode "The Big Salad," Jerry Seinfeld is really into his new girlfriend, Margaret—until he finds out that his archnemesis, his obnoxious and unattractive neighbor Newman, went out with her a few times. Jerry is especially horrified when he finds out that it was Newman who ended the relationship. He starts trying to find some previously unrecognized fault in Margaret. He's so obnoxious about it, though, that Margaret breaks up with him. This represents a crisis of existential proportions for Jerry. Because Newman broke up with Margaret, and Margaret broke up with Jerry, Newman appears to be Jerry's romantic superior.

If this episode doesn't illustrate mimetic desire clearly enough for you, then try "The Soul Mate" and "The Parking Space." But almost any episode of *Seinfeld* is shot through with the themes of mimetic theory—not because Jerry Seinfeld intended to, but because mimetic desire is a truth at the heart of human relations that Jerry Seinfeld and Larry David must have grasped intuitively to write such a show. The more accurately a work of art represents real human relations, the more it involves mimesis.[16]

Mimetic rivalries don't end well unless one of the two parties involved renounces the rivalry. To understand why, just imagine coming out on top of a rivalry. The act of winning paradoxically brings about defeat. It signals to us that we picked the wrong model in the first place. In the purported words of Groucho Marx: "I don't want to belong to any club that would accept me as one of its members."[17] And neither do we.

When one of the two parties to a rivalry renounces the rivalry, it defuses the other party's desire. In a mimetic rivalry, objects take on value *because* the rival wants them. If the rival suddenly stops wanting something, so do we. We go in search of something new.

Everyone has a toxic relationship to a model. The second half of this book is about the transformation of desire, which is the long-term cure. The short-term cure is to shield ourselves from infection.

CREATE BOUNDARIES WITH UNHEALTHY MODELS

You probably follow at least a few people who function as unhealthy models of desire for you. It might be an acquaintance or former colleague, someone you follow on social media, or maybe even a former classmate whose career you've followed through the years. You need to know what they're up to. You care what they think. You care what they want.

It's critical to distance yourself from the force they exert on you. Unfollow them. Don't ask about them. If you check up on them every day, then start by going at least a week before you check again. If you check on them every week, then go at least a month.

One of my friends was an early employee at a start-up in San Francisco when he found himself in a highly mimetic relationship with a talented colleague. The company grew so fast that there were a few months when they had to work nearly around the clock to keep up. If his rival Slacked the team to say that he was finally leaving the office at ten o'clock at night, my buddy would stay until ten-thirty the next night—and let everyone know about it. (It reminded me of my early days in investment banking, when none of the analysts dared to be the first one to leave for the day, lest anyone think they weren't working hard.)

It wasn't long before both my friend and his rival were pulling all-nighters. Not because the work demanded it, but because their mimetic rivalry did. Each one wanted to win the war.

Finally, my friend's rival left and became the eponymous founder of a company. Three months later, so did my friend. (He saw an "opportunity in the market," of course, around the exact time that his rival did.)

For months he followed the other guy's company and social media posts daily. He wouldn't admit to anyone, including himself, that his every move depended on what the other guy did.

When his rival bought Bitcoin, he had to buy Bitcoin, too, to make sure the other guy didn't hit a home run and leave him behind. My friend was like an investment manager who only buys index funds to ensure that he never falls behind the market, because that would be embarrassing. People do the same thing with their models.

When the Bitcoin bubble burst, my friend didn't care. As long as the other guy was wrong, he could be wrong, too.

Eight years have passed since they first struck out on their own. One day last year I ran across a news article profiling the rival and sent it to my friend. "Hey, look what Tony [not his real name] is up to," I wrote. To my surprise, my friend responded courteously: "Thanks for sending me this. I deleted it immediately.

About a year ago, I completely untethered myself from Tony to the point where I no longer even know what he's up to, and I'd like to keep it that way. Someday, once my rivalry with him runs out of oxygen and dies, I might not mind. But for now, I'm starving it to death. Can you do me a favor and not send me stuff like this?"

I was happy not to. And today my friend is happier, too.

Social Mediation

What we commonly call "social media" is more than media—it's mediation: thousands of people showing us what to want and coloring our perception of those things.

Tristan Harris, a former Google ethics executive and leader of the Center for Humane Technology, speaks about the danger of addictive design in tech. He claims that smartphones are like slot machines. Both work through the power of intermittent variable rewards—pulling the lever of a slot machine gives you a highly variable reward, which maximizes neurological addictiveness; your smartphone does the same thing every time you swipe down to refresh your Instagram feed, never knowing when something interesting might show up.

I respect Harris for being an advocate of human-centered design, but he misses a fundamental problem. Better design would help, but it only addresses part of the problem.

The danger is not that we have a slot machine in our pockets. The danger is that we have a dream machine in our pockets. Smartphones project the desires of billions of people to us through social media, Google searches, and restaurant and hotel reviews. The neurological addictiveness of smartphones is real; but our addiction to the desires of others, which smartphones give us unfettered access to, is the metaphysical threat.

Mimetic desire is the real engine of social media. Social media is *social mediation*—and it now brings nearly all of our models inside our personal world.

We live in Freshmanistan. Each of us has to examine what this means in our life—how mimetic desire manifests itself in the circumstances we're in, and how we should live.

This new world represents a threat but also an opportunity. Which new

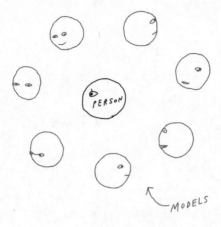

SOCIAL MEDIA

pathways of desire will emerge? Which new opportunities can we seize? How can we infect and be infected by desires that will ultimately lead to fulfillment and not to destruction? These are the questions that we'll finally have to ask and answer as individuals, and as a society.

Now let's see how mimetic desire works in groups.

Chapter 3

SOCIAL CONTAGION
Cycles of Desire

The Cigarette Test . . . Bullfighting Lessons . . .
Return on Collisions . . . Under the Anthill

> If individuals are naturally inclined to desire what their neighbors possess, or to desire what their neighbors even simply desire, this means that rivalry exists at the very heart of human social relations. This rivalry, if not thwarted, would permanently endanger harmony and even the survival of all human communities.
>
> —René Girard

In August 2019, two families went to a water park in California. By the time they left, they had gotten into a melee that left one man in a coma. Local news reported: "A large brawl broke out at Raging Waters in Sacramento, according to police. The melee, involving about 40 people, started as a dispute between two families over a beach towel."[1] A beach towel.

Few things are more mimetic than aggression. An argument breaks out between two people who each think a towel belongs to them. Within minutes, forty people are fighting over a single towel, behaving in exactly the same way, imitating one another in stupidity and violence.

This is essentially the plot of Shakespeare's *Romeo and Juliet*. It is not merely the tragic story of two young lovers. It's the tragedy of a warring city devolving into mimetic chaos. The opening line of the play is, "Two households, both alike in dignity." Yet they hate each other. The tiniest provocation has the potential to incite contagious violence that makes the families more alike, even as they think of themselves as more different.

As Peter Thiel points out in *Zero to One*, Karl Marx and William Shakespeare had very different views about why people fight. Marx thought conflict happens because people are different. People fight because they have different goals, desires, and ideas due to differences in the material goods they possess. In this framework, we would expect people who have the same material goods to fight less. Shakespeare's view seems to be exactly the opposite: people fight when they are similar, like the Capulets and Montagues in *Romeo and Juliet*.

The more that people in a group are alike, the more vulnerable they are to a single tension affecting the whole. Imagine the consequences of a single conflict in each of the following situations. In the first, you're walking down the street of a city when you see two random strangers in a brawl. In the second, a batter charges the mound in a game of Major League Baseball. In the first situation—two strangers fighting in the street—a couple of Good Samaritans might try to break up the fight. Most likely, no one else will get involved. But as every baseball fan knows, the fight between the pitcher and batter will probably clear the benches.

We'll see in this chapter that mimetic conflict is *contagious*. It can lead to a social environment in which everyone is reacting mimetically to everyone else. This dynamic keeps people locked in cycles of endless conflict, bound to one another through mimesis, unable to go anywhere.

Desire doesn't spread like information; it spreads like energy. It passes from person to person like the energy between people at a concert or political rally. This energy can lead to a cycle of positive desire, in which healthy desires gain momentum and lead to other healthy desires, uniting people in positive ways; or it can become a cycle of negative desire, in which mimetic rivalries lead to conflict and discord.

In Freshmanistan, the proximity and similarity of people make the stakes of mimetic desire higher. We'll spend most of our time there in this chapter and in the rest of this book.

We'll start in Italy, with a positive cycle of desire that led a tractor maker to build the world's first Lamborghini supercar. We'll also see how positive cycles of desire can lead to fitness, farming, and start-up success. We'll then go to downtown Las Vegas, where one entrepreneur attempted to build a city as though it were a start-up and accidentally launched a negative cycle of desire that led to mimetic chaos.

The different outcomes of these two stories are due to the way that desires were managed—or not managed—in each case.

Lamborghini versus Ferrari

Ferruccio Lamborghini first made his name by building tractors. He saw it as important and noble work for Italy's large population of farmers. But that all changed once he came into contact with Enzo Ferrari, whose beautiful cars he started driving after he became a successful businessman. As Lamborghini sat in his Ferrari, admiring its craftsmanship and power, something changed within him.

It had taken Lamborghini ten years to become one of Italy's most successful tractor makers. It would take only two years for him to become one of the most admired carmakers in the world. This is the untold part of the story—the story of desire.

THE STEALTH RACER

Northern Italy, sometime in the late 1950s. Ferruccio Lamborghini was on the Autostrada del Sole highway between Milan and Bologna in his red Ferrari 250 GTE Pinin Farina Coupé, blending in with traffic as he drove up and down the stretch of expressway where Ferrari technicians came to test their cars.[2] The nearby Ferrari factory didn't have a test track. So on the right morning, casual drivers might spy ten red Ferraris in their rearview mirrors before they roared past. The people driving those cars were among the best drivers in the world, pushing the vehicles to their technical limits.

Lamborghini lurked in traffic, waiting for the Ferrari test drivers. Once he spotted them, he swung out from the line of ordinary cars. His tires snaked side to side before gripping the pavement and launching the car forward, the acceleration pinning him back against the seat. Soon his older-model Ferrari joined the new, fresh-out-of-the-factory cars.

The Ferrari drivers weaved effortlessly in and out of traffic with expert clutch control, testing the torque and handling of the new machines. Lamborghini mingled among them. After a minute of toying with them, he pulled away from the pack. The other drivers chased him, but his Ferrari had found an extra ten miles per hour of speed.

Lamborghini, an accomplished mechanic, had made a few upgrades.

The region around Modena, where both Ferrari and Lamborghini are based, is the kind of place where everybody knows everybody else. The Ferrari test drivers knew who their rival was. The next time they saw him around town, fetching an espresso from his favorite café, they asked him, "Hey, Lamborghini, what have you done to your car?"

"Oh, I don't know," he said.

Lamborghini continued to taunt the Ferrari test drivers. At the same time, he kept having problems with his Ferrari, mechanical problems that seemed to occur far too often for such an expensive car. Even when the clutch was functioning, he wasn't satisfied with the way it felt when he shifted gears. It kept slipping.

A clutch on a manual transmission slips when it doesn't deliver power to the engine as designed. This usually happens when drivers shift improperly and wear out the disc that engages the engine. But Lamborghini knew what he was doing. The clutch wasn't slipping on account of him; it simply wasn't made well, or it was the wrong kind of clutch for such a powerful car.

The first few times he'd had problems with the clutch, he'd taken his car to the Ferrari factory, only to find the problem soon recurring. Exasperated, Lamborghini brought the car to the mechanics at his own factory. They discovered that Ferrari used the same clutch for its $87,000 luxury racing car that he used in his $650 tractors. Ferrari had been charging him a luxury markup every time they had to replace it, too. The car needed a bigger, stronger clutch. So Lamborghini replaced it with one of the best tractor clutches from his factory—and it solved the problem for good.

While he was at it, he decided to make additional improvements to the car's performance, including new cylinder heads with twin camshafts, which increased the airflow into the engine. The drivers of the new Ferraris didn't stand a chance against this tricked-out version.

Lamborghini took joy in driving his slip-free, souped-up model around the region, putting owners of basic Ferraris to shame with superior handling and speed. But that wasn't enough.

He had to let Ferrari know about his clutches.

* * *

Finally, at the start of the 1960s, Lamborghini had his chance to confront the maker of his vehicle.

"You know, today a guy came to me who has a tractor factory not far from here," Enzo Ferrari told his friend Gino Rancati, an automotive journalist for Radiotelevisione Italiana. "He explained to me that out of all the cars he owns, the clutch on the Ferrari slips more than any."[3] Ferrari was visibly displeased. He had repeatedly brushed Lamborghini off before he finally gave him an audience, viewing a tractor manufacturer as unworthy of his time. When they finally met, Lamborghini had the gall to patiently, maybe even condescendingly, explain the changes he had made to Ferrari's vehicle to improve it. There is no official account of how the meeting ended, but people who knew Lamborghini's account say that Enzo Ferrari could hardly contain his anger. In one telling of the story, Ferrari said: "The clutch is not the problem. The problem is you don't know how to drive a Ferrari and you break the clutch."[4] His implication: Lamborghini should stick to making tractors.

In the words of Ken Kesey, "it's the truth even if it didn't happen." Because whatever happened, Lamborghini left the meeting determined to build a superior car.[5] He knew that Ferrari had been ripping him off by marking up the price on the very same clutch he used in his tractors. And why did he have to beg for a meeting with a fellow manufacturer? Ferrari had disrespected him on several levels.

Until this point, Enzo Ferrari had been a remote or external model for Ferruccio Lamborghini, outside of his world. Lamborghini had seen his success on the racetrack and watched Ferrari's stature grow to legendary proportions. He was the clear G.O.A.T. of car manufacturing, and nobody dared compete with him. Ferrari had been a resident of Celebristan.

But now Lamborghini had come into direct contact with Ferrari. They were practically in each other's backyards, physically and socially. Lamborghini's factory was located seventeen miles away from Ferrari's. He, like Ferrari, had built a highly successful business. He was a millionaire who was driving, and improving, Ferrari's cars. What Ferrari wanted, Lamborghini began to want, too. Because of Ferrari's influence, he suddenly found himself wanting something he had never before wanted: to make the world's most beautiful, highest-performance supercar.[6]

A shift had occurred, and now both Lamborghini and Ferrari were living in Freshmanistan. Remember, Freshmanistan is defined by the *possibility of direct conflict*. Soccer stars Cristiano Ronaldo and Lionel Messi may be celebrities to most of us, but to each other they are not. The same was now true of Ferrari and Lamborghini. Through Lamborghini's success, they had come into close proximity and could compete directly.

LAMBO'S LEAP

In 1963, Lamborghini set up a new company called Automobili Lamborghini S.p.A. in Sant'Agata Bolognese, only a few miles from his tractor plant, on the outskirts of Modena.

The region was in the midst of a transformation. Emilia-Romagna had a long-standing reputation for bringing the world delicious prosciutto, parmesan, and balsamic vinegar. By the early 1960s, the area had also become the heart of luxury vehicle manufacturing in Italy. Maserati was headquartered in Modena proper, Ferrari was down the road in Maranello, and motorcycle company Ducati was in nearby Bologna.

Lamborghini now moved to poach the top engineers from the region's industrial companies and from his competitors. He offered them superior working conditions and benefits and the promise that he would put their skills to use making a car unlike any the world had ever seen. He pieced together a vision for his first car and his new factory from trips he took to factories in the United States and Japan, where he had studied their manufacturing processes so he could implement and improve them. "I don't invent anything," Lamborghini bragged. "I start where the others came from."[7]

Lamborghini introduced his first car to the public in 1964 at the Geneva Motor Show. The Lamborghini 350 GT was the first road car in history with a twelve-cylinder engine and a double camshaft. In 1966, Lamborghini introduced the Miura P400. It beat out Ferrari's best-performing road car in nearly every category of performance.

Three years after launching his car company, Lamborghini had produced a vehicle that dazzled even the most knowledgeable car enthusiasts. By 1968, just four years after making his first car, Lamborghini released the successor to the Miura P400, the Miura P400S, which became iconic. Frank Sinatra and Miles Davis each bought one. Eddie Van Halen can be

Tactic 4

USE IMITATION TO DRIVE INNOVATION

There's a false dichotomy between imitation and innovation.

They're part of the same process of discovery. Some of history's most creative geniuses started off by simply imitating the right model.

I sat down with Naresh Ramchandani, a partner at Pentagram, consistently ranked as one of the most innovative design firms in the world. They're the creative force behind projects such as the Harley-Davidson Museum, the set and on-screen graphics of *The Daily Show*, and the One Laptop per Child initiative.

"You can do innovation at any stage," Naresh tells me. "We sometimes start by saying, 'What's out there? What can we copy?'" The innovation comes at a later stage of the creative process.

If someone's primary objective is innovation for the sake of innovation, they usually end up in a mimetic rivalry with everyone in their field to compete primarily on the basis of originality. By devaluing all forms of imitation, they play a game of differentiation to get noticed. Being different for the sake of being different is the ethos behind shock-value art and academics whose salient feature is making outlandish claims to stand out from the pack.

As the fastest way to humility is not thinking more about humility but thinking less frequently of oneself, the safest route to innovation is also an indirect one. "There's great stuff out there," Ramchandani says. "Why wouldn't we learn from it? Why wouldn't we use it as an example, and build something on top of that rather than alongside it?"

Austin Kleon, author of *Steal Like an Artist*, put it this way: "If we're free from the burden of trying to be completely original, we can stop trying to make something out of nothing, and we can embrace influence instead of running away from it."[8]

Know when to lean into mimesis.

heard revving the engine of his P400S during his song "Panama." The original list price on these cars in 1968 was about $21,000 (nearly $170,000 in today's dollars). Today, they sell for close to $1 million.

Lamborghini's engineers—many of them hired away from Ferrari— were emboldened by the success of the Miura. They begged Lamborghini to be allowed to produce a true racing car and go head-to-head with Ferrari on the racetrack, confident that their engineering prowess would make them victorious.

Lamborghini wouldn't allow it.

BULLFIGHTING LESSONS

Ferruccio Lamborghini had a lifelong obsession with bullfighting. He understood its psychology.

In a bullfight, a bull is maneuvered into submission not by strength but by agility and psychology. The fight has three acts. In the first, the matador gets to know the behavior and quirks of the bull through a series of passes with a cape. In the second, the matador and his assistants stick sharp barbs in the bull's shoulders to wear him down. In the third act, called *muerte* (death), the matador kills the animal after bringing the bull to the point of physical and psychological exhaustion.

Being in a mimetic rivalry is like being a bull in a bullfight. In a bullfight, the matador orchestrates the actions of the bull. He makes the bull charge at a waving red cape, only to pull it away at the last second—just when the bull thinks he's going in for the kill.[9]

The bull is like Sisyphus, the mythological trickster. Zeus punishes him in the afterlife for his trickery: he makes him push a giant boulder up a mountain. Zeus enchants the boulder so that right before Sisyphus reaches the top, the rock slips away from him and rolls back down the mountain. Sisyphus has to go back to the bottom and start all over again—a task he must repeat for all eternity.

In a mimetic rivalry, a person's rival is like Zeus or the matador. The rival determines what a person wants next, which goals they pursue, what they think about when they go to bed at night. If a person doesn't realize what's happening, the game will bring them to the point of exhaustion, and maybe worse.

Ferrari gave Lamborghini the desire to make supercars. Lamborghini charged ahead. He became a formidable rival. But he refused to fight all the way to the end. He knew that there *was* no end. After all, the rivalry was never about cars. It was about honor.

Lamborghini didn't buy into the distortions caused by metaphysical desire, which leads people to seek satisfaction under a never-ending assortment of obstacles with no end. Girard explains the tragedy: "A man sets out to discover a treasure he believes is hidden under a stone," he writes in his first book, *Deceit, Desire, and the Novel: Self and Other in Literary Structure*. "He turns over stone after stone but finds nothing. He grows tired of such a futile undertaking but the treasure is too precious for him to give up. So he begins to look for a *stone which is too heavy to lift*—he places all his hopes in that stone and he will waste all his remaining strength on it."[10] Lamborghini chose not to.

"I refused to build it," Lamborghini said, referring to a race car. "Not just because I wanted to avoid fighting with Ferrari. It was a choice that concerned my role as a father. My son Tonino was sixteen when I started building cars, and I was sure he would be attracted to competition."

Lamborghini seemed to view competition as an occupational hazard for an entrepreneur—something good up to a certain point, but which devolves into rivalry if it's not kept in check. "This fear subsequently led me to include in the company charter a clause prohibiting participation in the [racing] wars," he added.[11]

Lamborghini took specific measures to mitigate the negative effects of rivalry. It saved him from the death of the bull.

In Tonino's official story about his father, he writes that Lamborghini enjoyed a peaceful final twenty years of his life at his vineyard, giving guests personal tours of his estate. Then he shares a juicy detail: his father always ended his tours by taking visitors to a plain-looking building near the main house. It could easily be mistaken for an abandoned barn. A small wooden sign hung near the main door. It read, 40 ANNI DELLA MIA VITA (40 Years of My Life).

Inside the barn was a collection of Lamborghini's most impressive creations: his rarest and best models of Lamborghini automobiles, tractors, engines, and parts. Lamborghini took his visitors through the barn, stopping at each exhibit, walking through the years of his life. But he always finished his tour of the barn with a demonstration: the cigarette test.

It went like this. Lamborghini would pop the hood of one of his cars and light a cigarette. He'd take the cigarette out of his mouth and place it directly on a cylinder head of the engine, instructing his visitor to keep their eyes on it. Then he hopped into the driver's seat of the car and put his foot on the gas until he had revved the engine all the way up to 6,000 RPM, forcing a tremendous amount of air through the intake valves—the equivalent of a thousand smokers taking lusty drags on the cigarette at the same time. The car roared, and the engine spun furiously. But the cigarette barely moved an inch as it was quickly consumed. The flawless mechanics of the car balanced thousands of moving parts without so much as a tiny rumble or shake—stasis through dynamics.

Lamborghini delighted in keeping up the act until the cigarette turned into a pile of ash. Then he would hop out of the car and sweep the ashes away with a wave of his hand.

Ferruccio Lamborghini died unexpectedly in 1993 at the age of seventy-six, but Automobili Lamborghini S.p.A. continues to this day. It ended 2019 with an all-time record in sales. And, eventually, Automobili Lamborghini entered the car racing business—the lure proved irresistible to future leaders. But it didn't happen on Lamborghini's watch. He knew when to apply the brakes at the right time and maneuver his energy into new opportunities.

Competition can be good up to a certain point. The key is knowing what that point is and having the wherewithal to pivot around it.

We're about to look at one project that suffered the consequences that Lamborghini avoided. But first, a quick look at how the spread of desire is different from the spread of information, and why it matters.

Memes and Mimetic Theory

How did tipping 20 percent become the norm in the United States but not in Europe? Why do Japanese businesspeople greet one another with bows instead of handshakes? Why do some organizations have well-developed lingo encoded in their handbooks and others don't? (And why is there so much lingo in the business world, period?) In all of these cases, imitation seems to play a large role.

In 1976, the evolutionary biologist Richard Dawkins coined the word "meme" in his book *The Selfish Gene*. He was attempting to explain the spread across time and space of nonmaterial things such as ideas, behaviors, and phrases. He called these things memes: cultural units of information that spread from person to person through a process of imitation.[12]

Dawkins's theory of memes and Girard's theory of mimetic desire both view imitation as foundational to human behavior. However, the two theories are different in almost every other respect.

According to Dawkins, memes work in a similar way to biological genes: their survival depends on their being passed on and replicated as perfectly as possible. They might mutate every once in a while. But in general, memes are discrete, static, and fixed.

According to meme theory, the spread of memes through imitation leads to the development and sustainability of culture. According to Girard's mimetic theory, culture is formed primarily through the imitation of desires, not things. And desires are not discrete, static, and fixed; they are open-ended, dynamic, and volatile.

We're all familiar with memes. They can be musical tunes ("Happy Birthday"), catchphrases (*studmuffin*), fashion (ties and high heels), even ideas ("What Happens in Vegas, Stays in Vegas"). Social media platforms like Twitter seem built to propagate them: words and ideas are spread through perfect imitation every time someone shares or retweets them.

Memes don't spread through human intentionality or creativity. As in Darwinian evolution, they undergo a series of random mutations and selections. (Internet memes, then, are *not* what Dawkins meant by the term *meme*—because an internet meme is a deliberate alteration of something.) True memes spread more like a virus. The individuals who spread memes are simply carriers—hosts through whom the information passes. Do you know who created the first lolcat? Neither do I. And it doesn't matter.

In Girard's mimetic theory, the opposite is true. People are not insignificant carriers of information; they are highly significant models of desire. We don't care about what is being modeled as much as we care about who is modeling it. We imitate not for the sake of imitation itself but for the sake of differentiating ourselves—to try to forge an identity relative to other people.

The drive to differentiate oneself is what we saw in the case of mirrored imitation, when someone does the opposite of what another person does (Seinfeld to Newman, hipsters to popular culture). Why do some people wear MAGA hats while others would not wear a MAGA hat if their lives depended on it? The revulsion that many people feel about wearing a hat that says MAKE AMERICA GREAT AGAIN (usually) has little to do with the color red, the style of the hat, or a political critique of the idea of national greatness. It has to do with the person modeling the hat: Trump.[13]

Most importantly, meme theory ignores all forms of negative imitation. In meme theory, imitation is, at worst, something neutral. From the standpoint of the memes themselves, it's something positive. In mimetic theory, imitation often has negative consequences. Because the imitation of desire causes people to compete for the same things, it easily leads to conflict.

In the rest of this chapter, we'll look at the flywheel effect of mimesis—the movement of creative and destructive cycles of desire, which are responsible for the rise and fall of cultures. They can't be captured in a meme.

The Flywheel Effect

Mimetic desire tends to move in one of two cycles. Cycle 1 is the negative cycle, in which mimetic desire leads to rivalry and conflict. This cycle runs on the false belief that other people have something that we don't have and that there isn't room for fulfillment of both their desires and ours. It comes from a mindset of scarcity, of fear, of anger.

Cycle 2 is the positive cycle in which mimetic desire unites people in a shared desire for some common good. It comes from a mindset of abundance and mutual giving. This type of cycle transforms the world. People want something that they couldn't imagine wanting before—and they help others go further, too.

In his book *Good to Great: Why Some Companies Make the Leap and Others Don't*, Jim Collins uses the example of a giant flywheel to explain how good companies break out and become great.

Collins asks us to imagine standing in front of "a massive metal disk mounted horizontally on an axle, about 30 feet in diameter, 2 feet thick, and weighing about 5,000 pounds," and that our goal is "to get the flywheel rotating on the axle as fast and as long as possible."[14] You push for hours, but the disk barely moves. Gravity is working against you. After three hours, you've achieved one full turn. Not discouraged, you continue pushing for a few more hours in the same direction with consistent effort. Suddenly, at an indiscernible point, momentum turns in your favor. The disk's weight is working *for* you rather than against you. The wheel propels itself forward. Five turns, fifty, one hundred.

Collins says this is what happens inside great companies when they put a positive, self-fulfilling cycle in motion. There is not a linear process of continuous improvement but a critical transition point at which momentum takes over and the process begins to power itself.

Mimesis, too, works like the flywheel. It accelerates in a nonlinear way—in both positive and negative cases.

The Creative Cycle

Giro Sport Design was founded by competitive cyclist Jim Gentes in 1985. It became one of the premier examples of the flywheel described by Jim Collins in his follow-up monograph, *Turning the Flywheel*.

When Gentes was in his twenties and working for a sporting equipment company, he spent his nights in his garage, tinkering with the prototype of a bike helmet that would change the landscape of competitive cycling. The helmet Gentes was working on weighed half of what other helmets weighed, and it had ventilation (helmets at the time had virtually none). His prototype was technically superior to any of the helmets that were in existence—by a long way. It looked great, too, which couldn't be said of any other helmet. The existing models were ugly appendages, necessary evils, an embarrassing half sphere of stuffy polycarbonate and foam.

Gentes took his prototype to the Long Beach bike show and received $100,000 in preorders. Serious cyclists could see from the start that the helmet was different. The validation from the bike show was encouraging. But he needed consistent order volume if he was going to quit his job and go all in.

By studying Nike, Gentes learned about the importance of social influence for athletic gear. If he could tap into the right influencer, he could reach a much larger network of loyal customers and gain the consistent order volume he needed.[15]

From his competitive cycling days, Gentes was friendly with the American cyclist Greg LeMond, who in 1986 had been the first non-European to ever win the Tour de France. LeMond had what Gentes was looking for: he was a strong rider known for taking risks, and he was handsome to boot.

LeMond had been seriously injured in a hunting accident in 1987, which caused him to miss the next two seasons. *Sports Illustrated* wrote glowing articles about his talent, even during his convalescence. Cycling enthusiasts were rooting for him to make a comeback at the 1989 Tour de France. It had been four years since Gentes had built the first prototype of a Giro helmet in his garage, and business was picking up. But he needed something to put him over the edge.

He approached LeMond about wearing his new Giro helmet, the first plastic-shell helmet on the market. Gentes promised LeMond that it would make him faster. He also used a significant portion of his company's capital to pay LeMond for the sponsorship, knowing his bet would pay off if LeMond was visible enough in the media's coverage of the race. And what if he won?

The outcome could not have been better. LeMond won the race—twenty-one days of racing over twenty-three days—by a grand total of eight seconds, the tightest finish in race history. Millions of people saw LeMond racing down hills in the French Alps wearing a helmet that looked to be half the size of other helmets, with sleek vents and bright colors compared to the dull turtle shells the other riders were wearing. The Giro brand's flywheel gained unstoppable momentum.

Giro's business flywheel, according to Collins, worked like this: "Invent great products; get elite athletes to use them; inspire Weekend Warriors to mimic their heroes; attract mainstream customers; and build brand power as more and more athletes use the products. But then, to maintain the 'cool' factor, set high prices and channel profits back into creating the next generation of great products that elite athletes want to use."[16]

Collins applies his flywheel concept to business growth. He shows that there are certain business models and processes that build momentum under the direction of a great leader. Like in a Rube Goldberg machine, one positive development triggers the next.

We can apply the flywheel concept just as well to the movement of desire. It's possible to set up our lives in such a way that we maximize the momentum of desire. Take fitness. For example: (1) I want to start working out, because my friend started a new workout program and looks great. (2) That makes me want to eat better, so that I don't negate my hard work at the gym. (3) So I want to turn down social invites that involve booze and Buffalo wings. (4) The result is that I want to go to the gym in the morning rather than pop Advil, slam coffee, and eat pancakes. (5) And that means I want to spend more time doing productive work. Eventually, I make wellness a virtue—meaning it becomes easy. Making healthy choices becomes something that I *want* to do instead of something I dread.

The fitness flywheel is hard to turn in the beginning. You feel like crap. Going to the gym seems daunting. And when you first start working out, it's painful. Change is imperceptible. But you keep pushing, and eventually the wheel starts turning. One day you wake up and look forward to your workout. The momentum has taken hold.

If you pick a spot on the outer edge of the flywheel and trace its movement from stage to stage, you are naturally pulled around the loop. Each step isn't merely the next step in a sequence; it's the logical consequence of the step that came before it. According to Collins, the movement of a flywheel works due to a *cannot help but* logic: you can't help but take the next step.

Giro followed this logic. If you make superior products, elite athletes can't help but want to wear them. If you get elite athletes to use your product, you can't help but attract the attention of mainstream consumers. And if you attract the attention of mainstream consumers, you can't help but build brand power. And when you have brand power, you can't help but increase your margins.

The flywheel effect plays out in both good and bad ways. Regenerative farms use positive flywheels. The farms are built around the health of their soil. The flywheel (I'm simplifying) works like this: Plants grow best in good soil, so you boost the biodiversity of the plants; which leads to healthier ruminants who eat the grass and the plants, and then poop; which leads to healthier soil; which means that water and beneficial microbes are retained better; which leads to even more nutrient-rich soil; which increases the vitality of the entire ecosystem.

But there are also negative flywheels, or "doom loops," where negative

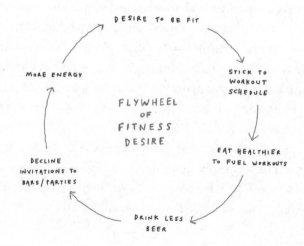

forces build on one another and lead to failure. A doom loop might work like this: An e-commerce company takes its focus off customer service to invest in other areas; which causes increased credit card chargebacks and bad website reviews; which leads to a decrease in order volume and returning customers; which causes a decrease in sales and inventory turnover; which forces the company to pay vendors late; which causes vendors to tighten credit terms and withhold inventory; which causes the company to focus even less on customer service because they are just trying to keep the lights on. Notice that the last of these stages leads right back to the first stage, amplifying the problem.

These positive and negative cycles play out in our lives every day. To add a layer of meaning to the concept of a flywheel—and to help us start positive ones—let's go back about 2,500 years to an insight that Aristotle had about a special force at work in living organisms and systems.

Aristotle invented the word "entelechy" to refer to a thing that has its own principle of development within it, a vital force that propels it forward to *become fully what it is.*

A human embryo, while dependent on others (especially the mother), already has within it a road map to develop into a fully formed human and the self-organization to get there, provided the embryo receives what it needs to support its growth. A standard computer doesn't possess entelechy: it has to be assembled and programmed. It can't assemble its own component parts and grow into a fully developed version of itself the way a sapling becomes a sequoia.

Understanding that some things have a vital principle of development and others don't is one way to understand the concept of a positive flywheel of desire: it contains the principle within it to help it achieve its purpose. Once you construct a flywheel and get it moving, it takes on a life of its own and begins to self-organize around an objective.[17]

Everyone has to construct their own flywheels. There is no one flywheel for fitness, for example; yours might look completely different than mine. The most effective personal flywheels come from people who know themselves well. You probably already have a tacit knowledge of what things increase and decrease the likelihood that you're going to want to do something in the future. The key is to make the cycle explicit, and then to put it in motion.

Tactic 5

START POSITIVE FLYWHEELS OF DESIRE

Desire is a path-dependent process. The choices we make today affect the things we'll want tomorrow. That's why it's important to map out, the best we can, the consequences of our actions on our future desires.

Start by thinking seriously about what a positive cycle of desire might look like for you. Start with a core desire. It might be spending more time with your kids, having more leisure time, or writing a book. Then map out a system of desire that makes it easier to bring that core desire to fulfillment.

Write it out. I suggest that each step in the flywheel be one sentence, contain the word "want" (or "desire"), and link to the next step in the process with a connector like *so that*, or *which leads to*, or *which makes*.

Here's an example from an e-commerce company that put a positive flywheel in motion for its customer service team, which had become complacent and unmotivated:

1. We want our customer service team to feel empowered to take ownership of decisions; *so that*

2. Customers feel they are speaking to someone with authority and therefore want to interact with them rather than asking for a manager; *so that*

3. Efficiencies are created that allow managers to spend less time talking to frustrated customers and more time managing projects they want to be working on; *so that*

4. We can create a discretionary bonus pool, administered by managers who want to reward customer service team members who take ownership of decisions; *so that*

5. The customer service team members want to take more ownership of decisions.

Yours doesn't need to have five steps. But make sure that each step leads inevitably to the next, and that the last step in the process leads back to the first.

Negative flywheels are far more common than positive ones. This is especially the case in Freshmanistan, where people have more in common and they are in close proximity. Like the warm ocean water over which hurricanes form, mimetic contagion is able to gather steam faster in Freshmanistan because everyone is in a reflexive environment, picking up on mimetic cues.

I became caught in a negative flywheel during the time I spent in downtown Las Vegas exploring the culture of the online shoe seller Zappos.com. I aspired to many of the things that its CEO, Tony Hsieh, modeled: simplicity, a flat organizational structure, a willingness to do things differently. He even modeled weirdness and enshrined it as a value in his company. But I didn't know at the time how mimetic desire works. Neither, apparently, did anyone at Zappos.

The Destructive Cycle

Tony Hsieh wanted everyone to be happy.

"Are you happy?" he asked me one day as we were getting to know each other. Zappos was the hottest story in business media at the time. Everyone was fascinated with the company culture. It was portrayed as something like Willy Wonka's chocolate factory, with Tony as Wonka—the zany, mega-wealthy founder who gave tours of his operation to people curious about how he'd built a utopia of happiness.

Serena Williams had come by the Zappos HQ for a tour and conversed with Tony. I once wondered if the company had a world-class publicist; it turns out that they simply delivered excellent customer service year after year, and the effort eventually paid off. The right people noticed.

When the company first started out, its operational flywheel looked something like the illustration below.

Nick Swinmurn, who founded Zappos in 1999, simply wanted to make shoe buying easier. He saw an opportunity to make a painful process easy with the birth of e-commerce.[18] In order for the business model to work, the company needed to boost its sales and customer base—especially its

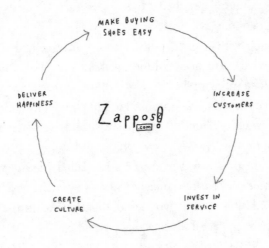

customer retention (the percentage of customers who come back and buy more stuff), which was the key to becoming profitable.[19] Tony Hsieh started off as an investor in the company and eventually became its CEO. In early 2003, he and Fred Mossler, one of the first employees, realized that customer service should be a key focus.[20] Then, in 2004, they recognized that the focus on service could only be accomplished by focusing on company culture. Eventually, they discovered that the organizing principle of their culture was "delivering happiness."

The culture would revolve around delivering happiness to all stakeholders: employees, investors, vendors, and more. Happy people make the entire flywheel turn more easily. In 2008, the company surpassed $1 billion in sales at least two years ahead of schedule.

This success came in large part due to Tony's leadership. He was passionate about the work that Zappos did, he was willing to take huge risks to get the company off the ground, and he made Zappos a place where people genuinely wanted to work.

But when I look back, I don't believe anybody knew about or took into account the negative effects of mimesis, which effectively hijacked a positive flywheel of desire that had made Zappos a profitable company and a fun place to work. The change came subtly and imperceptibly. Nobody felt there was a problem—like frogs sitting in a pot of gradually warming water who never feel the urge to jump out before they become frog soup.

Back to my conversation with Tony.

I told him that I was happy.

"Are you, though?" he asked.

"Yeah. I'm good."

Tony smiled with his eyes only. He was an excellent poker player. I didn't know what hand he was holding this time. "But . . . are you *really* happy?" he asked me again.

I think that's when I was supposed to fling myself into his arms crying. Instead, I threw up my hands. "I think so!" I said, this time more irritated and unsure of myself. "Why?"

Tony told me that he was reading a book called *The Happiness Hypothesis: Finding Modern Truth in Ancient Wisdom* by social psychologist Jonathan Haidt. Did I know, Tony wondered, that the one thing that all people seek, everywhere and at all times, is happiness?

Tony's logic went like this: businesses must exist to make customers happy; therefore, the more we can learn about the science of happiness, the more effectively we can build a successful business.

At least that was the idea.

A year later, Tony sold Zappos to Amazon for around $1.2 billion.[21]

Shortly after the sale, he wrote a book called *Delivering Happiness: A Path to Profits, Passion, and Purpose*. He also announced the launch of the Downtown Project, a roughly $350 million investment in downtown Las Vegas. Tony planned to take the culture of happiness that he'd helped build at Zappos and use it to help a city.

The goal of the Downtown Project was to revitalize the area north of Fremont Street, a run-down stretch known more for opioid addicts and prostitution than gambling. It was the end of the earth for gamblers, a part of downtown Las Vegas to which no tourist ever ventured unless they were on their fifteenth frozen daiquiri.

Between 2010 and 2013, Tony Hsieh and his partners spent about $93 million to buy up twenty-eight acres of land and buildings, from vacant hotels to high-rise condos and struggling bars. Their long-term goal was to invest in the space and spread the Zappos culture to its inhabitants, lure talented entrepreneurs from Silicon Valley, and ultimately create an entrepreneur-driven ecosystem.

It was a social experiment—"the city as a start-up," Tony called it. A happy city.

After the sale to Amazon, Tony stayed and remained the leader of Zappos. It was allowed to operate as a largely autonomous company. At the same time, Tony launched the Downtown Project. Its culture blended into the Zappos culture, and vice versa. Zappos and the Downtown Project were part of the same ecosystem.

There were warning signs from the beginning. Zappos employees told me that morale was low. There were too many changes, too fast—including the adoption of an experimental flat management structure. It created chaos.

The Downtown Project was in the same situation. According to Nellie Bowles in her 2014 exposé in *Vox*, less than a year into the project, Jody Sherman, one of its star entrepreneurs, shot himself in his car.[22]

A year after Sherman's death, Ovik Banerjee, a key member of the

Downtown Project and of Las Vegas's first Venture for America group (Venture for America is Andrew Yang's nonprofit organization to train young start-up workers and founders), jumped from his high-rise apartment terrace downtown.

Less than five months after Ovik's death, Matt Berman, founder of Bolt Barbers, one of the start-ups incubated by the project, was found hanging in his room.

MOVING TO FRESHMANISTAN

What went wrong in downtown Las Vegas? The new flat management structure of Zappos—and by extension the Downtown Project—had mimetic consequences that nobody had foreseen or taken into account.

Ovik Banerjee "never had a clear job," according to a source quoted by Bowles in her Vox piece. "*No one* had a clear job," the source continued (emphasis mine). "It was Tony leading people saying, 'Come out and do things, come out and have fun,' and then you get here, and there's no structure."

When the focus shifted away from shoes and customer service to happiness, the number of mimetic models multiplied. It was unclear who was happy and who was not; whom to imitate and whom not to imitate; who was a model and who was not. Zappos and the Downtown Project had turned into Freshmanistan.

Dr. Zubin Damania, who ran a health clinic in downtown Vegas and was a part of the Downtown Project team, commented to Bowles on the situation: "In an entrepreneurial community, so, so many boundaries are gone. People are taken away from their social moorings. It's extremely high-pressure."

He suggested that the illusion of freedom—the idea that every entrepreneur is a master of their own desire—is dangerous. "Founders are the worst," he said. "There's a Randian—I must be the John Galt—feeling. You can be as liberated as you want, but there's a web of connectivity, and they forget."

Desire is part of the web of connectivity. When people deny that they are affected by what other people around them want, they are most susceptible to getting drawn into an unhealthy cycle of desire that they don't even know to resist.

Mimetic desire breeds rivalries, which breed collisions and conflict.

Every community in a mimetic crisis—that is, every community that suffers a loss of difference, where there is no clear separation between

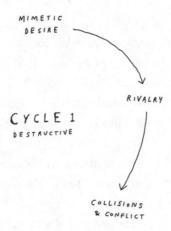

MIMETIC
DESIRE

RIVALRY

CYCLE 1
DESTRUCTIVE

COLLISIONS
& CONFLICT

models and imitators—has its own version of the Cycle 2 flywheel. In the Downtown Project, there was an intentional focus on making collisions of people happen. Without knowing it, they were exacerbating mimetic rivalry.

Tony liked to use a key metric to measure success, which he calls "return on collisions" (as opposed to return on investment). According to Tony, a "collision" is an unexpected, serendipitous meeting between two people that leads to positive outcomes. For instance, two entrepreneurs work next to each other at a coffee shop and eventually strike up a partnership, or investors find new people to invest in over gin and tonics at a bar. In Tony's eyes, return on collisions was the best way to measure how culture, or community, was leading to value creation.

"My fascination with serendipity started in college," Tony said, according to a 2013 *Inc.* article. "I think for most people, college was the last time it was normal to just randomly run into people all the time. As you get older, you drive to work, see the same people every day, then go home. But the best things happen when people are running into each other and sharing ideas."[23]

He wanted to make downtown Las Vegas more like college. A Freshmanistan.

But not all collisions are alike. Some lead to good things: friendships, marriages, and ideas for new companies. Others lead to chaos and confusion.

RETURN ON COLLISIONS

One of Tony's strategies for encouraging collisions was the optimal use of space. He wanted to maximize the occurrence of random encounters. He and his colleagues would stage concerts and meet-ups, hackathons, happy hours, open-mic nights, and an open-door feel at the Ogden, a high-rise in downtown Vegas where many Zappos executives lived. The atmosphere was something like a college dormitory, where closed doors are frowned upon and anyone can pop into anyone else's room at any time.

I attended a collision-focused meet-up one night for the Downtown Project in Tony's penthouse. We were in a large room with hardwood floors and floor-to-ceiling windows that looked down onto the Fremont Street Experience. The only furniture in the room was a couple of dozen rolling desks with attached seats, the kind you might find in a third-grade class-room where the teacher is asking the kids to move around a lot and form groups. These were adult-sized desks, but the idea was the same. There was one desk for each person.

Tony walked into the room and, hands in pockets, told everyone to take a seat. He explained that we should try to meet as many people as possible in the next hour—like speed-dating for start-ups. We were to roll around the room and "collide" with one another, then strike up a conver-sation. He took a seat at one of the desks himself and rolled around with the rest of us.

I don't remember anybody I met that day. I do know that when the experience was over, I left the room with more anxiety than I'd had when I sat down. I was now comparing myself to at least twenty other ambitious people, most of whom lived within the same four-block radius or planned to move to downtown Vegas if they didn't live there already. Forget about the rolling desks—we were all sitting in invisible, ego-driven adult bumper cars. The collisions were happening faster and harder by the day.

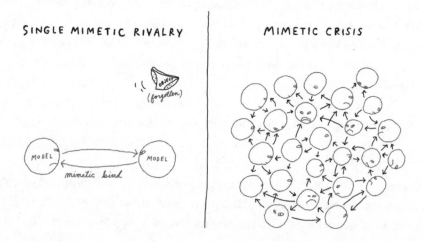

SINGLE MIMETIC RIVALRY   MIMETIC CRISIS

UNDER THE ANTHILL

The Downtown Project was an extension of Zappos, and Zappos had always been known as a relatively flat organization—there were few layers of management between executives and staff. But Tony wanted to take things a step further.

In 2013, Zappos implemented a new management philosophy called "holacracy." Brian J. Robertson, the founder of Ternary Software, helped develop holacracy as a "social technology" and an "operating system" for organizations because he was looking for a better way to run his own company. Robertson trademarked the term and worked hard to implement it in his own company and test it in others. Soon large companies such as Zappos and online publishing platform Medium adopted holacracy as well.

In his book *Holacracy: The New Management System for a Rapidly Changing World*, Robertson describes how Tony approached him after hearing him speak at a conference. "Zappos is growing," Tony told him. "We've reached fifteen hundred employees, and we need to scale without losing our entrepreneurial culture or getting bogged down in bureaucracy. So I'm trying to find a way to run Zappos more like a city."[24] Holacracy was a codified system to help him do it.

Holacracy replaces traditional management hierarchies with self-organizing teams of people working on a specific project. The idea is that traditional titles such as CEO or COO are eliminated in favor of a number of roles that serve the same organizational purpose, and different people can fill these roles at different times; they are elected to fill those roles through a governance process, according to a constitution.

Part of the reason for doing this, Robertson told me when we spoke, is to differentiate a *person* from a *role* for the sake of making the best decision for the organization. This helps remove personal ego from mission-critical processes. But the decoupling of role from person can sometimes bring hidden problems to light.

As part of the transition to the new system, Tony stepped down as CEO. The previous management hierarchy of Zappos disappeared almost overnight. The Downtown Project made a similar move to holacracy. What happened, as we'll see, was a mimetic crisis—not necessarily a failed management system, as the popular business press has painted it, but a system that opened the door for hidden mimetic desire to boil over.[25]

A human-centered approach to business involves grappling with the messiness of human interactions—with human nature. To introduce something foreign to human nature, which doesn't complement it—like an organizational "operating system" that doesn't account for mimetic desire—is to open up a Pandora's box.

Zappos had eliminated the management hierarchy, but they couldn't eliminate the network of desire and the need that people have to be in relationship to models. There is always a hierarchy of desire from the perspective of an individual: some models are worth following more than others, and some things are worth wanting more than others. We are hierarchical creatures. This is why we like listicles and ratings so much. We have a need to know how things stack up, how things fit together. To remove all semblance of hierarchy is detrimental to this fundamental need.

When Zappos moved to holacracy, what disappeared aboveground— the visible roles and titles—reappeared in different ways underground.[26] "The environment became more political," journalist Aimee Groth, who wrote about holacracy for *Quartz*, told me. "People were less secure in their jobs . . . less clear on how they could hold on to their roles and their jobs. However, you still had a few people who had infinite power because they had a strong relationship with Tony." There was a hidden web of desire that nobody could decipher.

Unwittingly, the gates of mimetic rivalry had been flung wide open. By 2010, when Tony published his book *Delivering Happiness*, the company's flywheel seemed to have a new starting point. In the original one, "delivering happiness" was the *last* stage in the process. In the new version, happiness seeking became the *first and primary* step in the process.

It's presumptuous for anyone to think that they can "deliver happiness" to anyone else—even their own spouse. It's not our job. It's certainly not the job of a company.

The mission "delivering happiness" is radically different from the mission "make buying shoes easy" as the starting point of a flywheel. It's more ambitious and meaningful, but it's also more dangerous.

Most people gauge their happiness relative to other people. When the starting point of a flywheel is the delivery of happiness—both for custom-

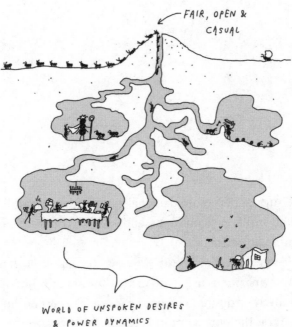

THE "NON-HIERARCHICAL" COMPANY

FAIR, OPEN & CASUAL

WORLD OF UNSPOKEN DESIRES & POWER DYNAMICS

ers and for a company culture—the system revolves around a vague notion of happiness that is rife with mimesis.

When happiness becomes the dominant desire in a community in which nobody knows what happiness is or how to achieve it, everyone looks to their right and their left to find models of desire that seem worthy of following. And since everybody in Freshmanistan is in close proximity and on a level playing field, it leads to a conflict of all against all.

I saw happiness treated like a meme—something that could be delivered or transmitted from person to person by following a formula. But happiness is not a meme, and it can't be delivered.

People always pursue happiness by looking for models of happiness— whether that is someone who has lived the American dream, a Silicon Valley CEO, or your next-door neighbor. But external hierarchies are merely

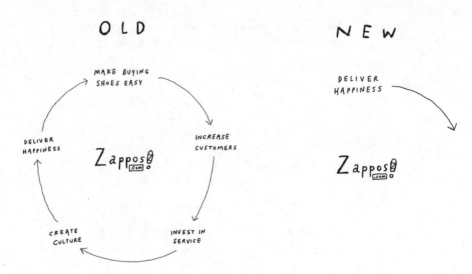

the visible surface of a more personal system: the structure of desire that lives invisibly inside each one of us, and which is connected to other people through mimetic desire.

C. S. Lewis called this invisible system the *inner ring*. It means that no matter where a person is in life, no matter how wealthy or popular a person is, there is always a desire to be on the inside of a certain ring and a terror of being left on the outside of it. "This desire [to be in the inner ring] is one of the great permanent mainsprings of human action," Lewis said. "It is one of the factors which go to make up the world as we know it—this whole pell-mell of struggle, competition, confusion, graft, disappointment and advertisement. . . . As long as you are governed by that desire, you will never get what you want."[27]

Zappos dismantled any visible signs of an outer ring. They forgot about the inner one.

Hierarchical Values

Tony's project to make downtown Las Vegas an entrepreneurial hub and happy community was a noble one, in principle. Its downfall was an impoverished view of human nature.

CEOs, teachers, policymakers, and others responsible for shaping an environment should understand how decisions affect people's desires. As a city planner needs to consider the effect of parks and murals and bike paths on everything from traffic to crime, so a good leader needs to consider the impact of their decisions on *human ecology*—the web of relationships that affect human life and development. No aspect of human ecology is more overlooked than mimetic desire.

Early on at one of my companies, I made the mistake of forming a way-too-serious flag football team that competed in a city league, not realizing that it divided our young start-up into factions. Having fun and freely associating outside of work was not a problem. The problem was that I, the CEO, was the one who organized and led the effort. At that stage of our company (there were only about ten of us), the idea and organization needed to come from someone other than me for it not to feel like a top-down imposition of cultural expectations. My football fanaticism inflamed a few rivalries and bent desires toward small-spirited goals rather than great ones.

Leaders should also consider that economic incentives are always more than economic. If the signals are strong enough, they can distort desires and give people a "false north" on their career compass. Imagine a university that gives history majors $10,000 in cash, but nothing for any other major. The cash would introduce an anomaly into the market for majors. It would shock no one if a few students suddenly realized that they "wanted" to be history majors. They might really convince themselves that their newfound course of study was an expression of their authentic selves.

You know by now how tenuous desire is. Yet we do things like this all of the time. Parents will invest heavily in equipment and training for a child in one sport but not another, or offer to pay for one college (or one specific course of study) and not another. The children don't always have the psychological freedom or maturity to separate what they want from what is expedient.

Some people wake up one day after twenty years and wonder how they ever got into their career and why they are still there. You can influence behavior with financial incentives, no doubt—but economic incentives alone don't explain why people are captivated by certain models. You can't buy desire. When we subsidize risk, we are left with a distorted view of who wants what. Sometimes, that person is us.

Marketing, money, and models distort desire for people if there is no clear hierarchy of values. This phenomenon is apparent with my freshman business school students. They're in a formative, hyper-mimetic stage of their lives. Because I know that I'm a model for many of them, I'm careful about signaling value for certain majors, internships, or jobs—at least until I get to know them. I grow frustrated when students come to me about three weeks into the semester saying they got the full-court press from an aunt or uncle or friend or career counselor about the job security that awaits them if they major in accounting. (I'm picking on accounting here—I could have chosen anything.) Even students who don't have an accounting bone in their body begin second-guessing the track they're on. And it's not their fault: the person modeling a career in accounting is an excellent model. They exude tranquility, they're financially stable, they seem happy.

"Is accounting what you truly want to do?" I ask.

"I . . . don't know," they say. "Maybe?"

It's as if they're at a buffet in a foreign country filled with hundreds of foods they've never tried and maybe never knew existed. The first thing they're going to do is get in line behind someone who looks like they know what they're doing. We all do it.

I talk to them about what their hierarchy of values might look like. What's important to them? What are the five, or ten, or twelve rules they want to live by?

It's not enough for them to tell me all the things they care about. They have to put them in some kind of order. I challenge them to make decisions and prioritize them. Augustine of Hippo called this the *ordo amoris*, or the "order of love."[28]

Values and desires are not the same thing. Values act to order desires the way they do a diet. If a person who loves meat realizes that their values are such that they no longer wish to eat meat, there comes a time—after they've lived out their values long enough—when they no longer even *want* meat. You could put the juiciest burger imaginable in front of them and they wouldn't be tempted to eat it.

For many people, ordering desires starts out unconsciously. It may be this simple: I take care of my immediate family first, then others; I answer emails from people I know first and respond to unsolicited sales inquiries later; or if I only have time to clean one room in the house today, I clean the kitchen.

Whether we recognize it or not, our minds think in hierarchies all of the time—whether it's related to our daily to-do list, the priority of issues in an election, or even a glance at a menu in a restaurant (appetizers, main course, dessert). Without a hierarchy of values, which helps form and direct desires, we can't even begin to think about what to pay attention to and to what degree.

Most companies have a mission statement. Many of them have core values or some equivalent. Few make explicit—either to themselves or to the outside world—their hierarchy of values. This makes it hard for them to determine what to do when two of their values come into apparent conflict, like the value of protecting the health and safety of all employees versus keeping the business running during the COVID-19 pandemic.

A hierarchy of values is especially critical when choices have to be made between *good things*. If values are all equally important, or if there isn't a clear understanding of how they relate to one another, mimesis becomes the primary driver of decision-making.

"My friends and my faith are both super important to me," one of my college buddies says. Good. But what will he do if one of his best friends schedules his bachelor party in Miami's South Beach on a high holy day? Saying that two different things are "super important" won't help. Without a clear hierarchy, he's more likely to choose according to the influences around him. His decision will be mimetically driven, not values driven.

Companies face situations where competing claims are made on their values every day. Two of a company's core values might be "inclusiveness and diversity" and "relationships built on trust." If the company is in an industry where sales happen within an old-boys' club, where trust has already been established between its members, how can they hire a young woman for a sales role and give her a chance to build up trust in a different way if there is not a clear priority on inclusiveness and diversity in the hiring process? In the absence of a hierarchy of values, the hiring managers won't know what to do when 95 percent of the applications they receive are from highly experienced men. Mimetic forces will continue to dominate their sales force.

In a company's capital structure, there is always a hierarchy of claims on the company's money. A start-up's cap table (a hierarchical list of who owns what and who gets paid what and when) might contain the following groups of people with their claims listed in order of priority:

Tactic 6

ESTABLISH AND COMMUNICATE
A CLEAR HIERARCHY OF VALUES

A hierarchy of values is an antidote to mimetic conformity. If all values are treated as equal, then the one that wins out—especially at a time of crisis—is the one that is most mimetic. (During the early stages of the COVID-19 outbreak, there was panic-buying of toilet paper. There was no issue with supply; there was an issue with mimesis. Cultural values are often subject to the same irrationality—people tend to panic-buy whatever is most important to them at the moment rather than what is best for the common good.)

It's not enough to name values. They need to be ranked. When all values are the same, nothing is being valued at all. It's like highlighting every single word in a book.

It's better to construct a mental model of your hierarchy of values (or your shared values, if you are in a relationship). Map it out on paper. Encourage your company to do the same. That hierarchy can change as time goes on. But by stratifying your values, you will be able to weigh and measure options when you have to make decisions in complex situations.

Remember that conflict is caused by sameness, not by difference. If everything is equally good or important, the propensity for conflict is higher. Don't contribute to the tyranny of relativism. It has too many tyrants as it is.

A lack of clear, prioritized values in many companies allows mimesis to hijack the concept of corporate social responsibility (CSR) and turn it into a flaccid marketing gimmick. It's not that the values espoused by CSR programs aren't important. But one gets the feeling that even "social responsibility" has become a mimetic, virtue-signaling game—more "social" than responsible.[29] Avoid that by establishing and communicating both your values and their relative importance.

Some values are absolutes. Know them. Name them. Defend them. They make up the base of the pyramid, or the center of your concentric circles (depending on how you choose to depict the hierarchy).[30]

secured creditor, unsecured creditor, preferred stockholders, common stock series A holders, common stock series B holders, and founder's stock.[31] If we demand such a clear hierarchy when it comes to who gets paid first, coming up with a similar hierarchy for our values—indicating what we should desire first—is the least we can do.

The Collapse of Desire

Value systems with a clear hierarchy are more effective during crises than systems of values that lack a hierarchy.

Ferruccio Lamborghini had such a hierarchy. Protecting his son from a life of rivalry on the racetrack, and the accompanying potential for injury or death, was more important than winning the game at all costs. When mimetic escalation reached a frenzied state, Lamborghini did something seldom done during a time of peak desire: he quit. He could do so because he was checking his desires against a clear hierarchy of values, which prevented them from spinning out of control.

People don't think about the costs of a car crash until after they've been in one. Practically nobody thinks about collisions of desire before they happen. For Lamborghini, they were the same thing: colliding desires meant colliding cars.

Tony Hsieh wanted to maximize positive collisions, but he didn't take into account the hidden collisions of desire that happen in the mimetic space between people, in the hidden recesses of the human heart.

On August 24, 2020, the *Las Vegas Review-Journal* confirmed that Tony Hsieh was stepping away from Zappos after more than twenty years at the helm.[32] I learned on November 27, 2020, a few short months before this book went to print, that Tony had died the day after Thanksgiving—almost twelve years exactly since Tony and I shared a memorable Thanksgiving dinner together at his home. I am in awe of all that he accomplished in his 46 years. The Downtown Project endures. There are many admirable projects that have sprung from it, such as the Writer's Block, an independent bookseller, and Chef Natalie Young's restaurants.[33] Still, between 2015 and 2019, much of the media coverage about the Downtown Project was negative. Some of that criticism was just. But accounts that remain on the surface—"management theories"—never tell the whole truth.

In downtown Las Vegas, where there was confusion about who was a model to whom, one model stood out from the rest: Tony himself. He was extraordinarily wealthy, yet it was common to see him at a hole-in-the-wall restaurant or bar in the Downtown Project's radius, every bit as approachable as the gamblers drinking Bud Light at the slot machines a few blocks up Fremont Street.

He was a paradoxical figure. Tony genuinely wanted other people to be happy, but he didn't seem to desire anything for himself. As the contagion of desire spread in downtown Vegas, Tony stood out from the rest. As we'll see in the next chapter, that's a huge risk.

Do you want to know what the Egyptians did to their cats?

In 2018, a sarcophagus from ancient Egypt was discovered with dozens of mummified kittens in it. Findings like this—which go back to at least 1799—have dispelled the myth that the Egyptians were the ultimate cat lovers. The truth is darker.

The Egyptians used their cats for ritual offerings and sacrifice. That's *why* they were considered sacred. In mimetic theory, there is a near-indissoluble link between chaos and order, violence and the sacred. Sacrificial rites—whether sacrificing cats in ancient Egypt or the ritual firing of coaches and CEOs today—are the mechanism by which mimetic contagion is contained and controlled.

The fourth and final stage of the mimetic cycle, which we turn to now, is the process by which chaotic desires become orderly desires in human societies: the scapegoat mechanism.

Chapter 4

THE INVENTION OF BLAME
An Underrated Social Discovery

The Danger of Purity . . . Safety in Judgment . . .
Ritual Scapegoats . . . Self-Awareness, Self-Hatred

> I wonder if some aspect of human nature evolved in the context of competing packs. We might be genetically wired to be vulnerable to the lure of the mob. . . . What's to stop an online mass of anonymous but connected people from suddenly turning into a mean mob, just like masses of people have time and time again in the history of every human culture?
>
> —Jaron Lanier, computer scientist and philosopher

Between 1977 and 1982, Jenny Holzer stalked the streets of New York City under the cover of night, pasting her subversive artwork on walls. She called them "Inflammatory Essays"—lithographs on colorful pieces of paper, each one with exactly one hundred words printed in italicized, capital letters, left-justified, twenty lines.

The words were taken from literature and philosophy, from anarchists and activists and extremists. The first five lines of one said:

DISASTER DRAWS PEOPLE LIKE FLIES.
SPECTATORS GET CHILLS BY IDENTIFYING
WITH THE VICTIMS, FEELING IMMUNE ALL
THE WHILE! THIS IS A PARTICULARLY
UNATTRACTIVE FORM OF VOYEURISM.[1]

In the mid-1980s, Holzer took her art to an eight-hundred-square-foot Spectacolor light board in Times Square as part of the Public Art Fund's "Messages to the Public" program. In 1982, Times Square was already a light show of advertisements, the seedy center of New York City tourism— and of American consumer culture. Holzer lit up the massive board with 250 works from her "Truisms" series. The words were formed with white LED lights on a black background. One of the statements said:

> PROTECT ME
> FROM WHAT
> I WANT

The message stood in striking contrast to the frenetic color, movement, and noise of its surroundings. Holzer's admonition made hurried souls stop and ponder its meaning.

Her plea to be protected from what she wants is something that everyone can relate to. Each of us has desires that, if followed to the end, are dangerous to ourselves and to others. The same is true at the societal level: out-of-control mimesis makes desires spread and collide violently.

René Girard saw that for thousands of years humans have had a specific way of protecting themselves in a mimetic crisis: they converge, mimetically, on one person or group, whom they expel or eliminate. This has the effect of uniting them while providing an outlet for their violence. They protect themselves from what they want—from their mimetic desires, which have brought them into conflict with one another—by directing their desire to vanquish their rivals to a single fixed point: someone that has become a proxy for all of their enemies. Someone who is unable to fight back. A scapegoat.

Sacred Violence

Girard saw a close connection between mimetic desire and violence. "People everywhere today are exposed to a contagion of violence that perpetuates cycles of vengeance," he said in his book *The One by Whom Scandal Comes*. "These interlocking episodes resemble each other, quite obviously, because they all imitate each other."[2]

How do these cycles of vengeance start? Mimetic desire. "More and more, it seems to me," wrote Girard in the same book, "modern individualism assumes the form of a desperate denial of the fact that, through mimetic desire, each of us seeks to impose his will upon his fellow man, whom he professes to love but more often despises."[3] These small, interpersonal conflicts are a microcosm of the instability that threatens the entire world. And before the world: our families, cities, institutions.

The nineteenth-century Prussian general and military theorist Carl von Clausewitz wrote a book called *On War*, mandatory reading in many military schools. Girard credits him with having recognized the mimetic escalation of most conflicts. Early in *On War*, von Clausewitz wonders: "What is war?" His answer, which he takes the rest of the book to explain: "War is nothing but a duel on a larger scale."[4]

War is the escalation of mimetic rivalry. Where does it end?

Throughout most of human history, there were clear winners and losers in war, recognized as such through formal processes. Conflicts came to an end when one side admitted defeat according to rituals such as the signing of a peace accord. Not so today, when terror cells can spring up from within a community and then grow like a hydra when any of their members are struck. How could there ever be a definitive conclusion to a war in which combatants masquerade as ordinary citizens? Girard thought we had entered a dangerous new phase of history, ripe for what von Clausewitz called "the escalation to extremes"—the desire of each side in a conflict to destroy the other, which reinforces and escalates the desire of the other for violence.

It seems that even in Clausewitz's time—and certainly by World War I—war had already become an escalation to extremes. Something had shifted. There were no more buffers or brakes on how far and wide the destruction of war could go. Today, we see the escalation of extremes in political rhetoric and positions. And for the first time in human history, we have the technological means to destroy ourselves. It's unclear what mechanisms will be effective in stopping the escalation.

This is very different from earlier societies, which used a terrifying social innovation to contain the spread of conflict.

In his study of history, Girard found that humans time and time again turned to sacrifice in order to stop the spread of mimetic conflict.[5] When societies were threatened with disorder, they used *violence to drive out violence.* They would expel or destroy a chosen person or group, and this action would have the effect of preventing more widespread violence. Girard called the process by which this happens the *scapegoat mechanism.*

The scapegoat mechanism, he found, turns a war of all against all into a war of all against one. It brings temporary peace as people forget their mimetic conflicts for a while, having just discharged all of their anger onto a scapegoat.

This process, Girard believed, was the foundation of all culture. The institutions and cultural norms that we find around us, especially sacred rituals like elections and capital punishment, as well as many taboos, are mechanisms that were developed to contain violence.

In this chapter, we'll see how the scapegoat mechanism is still at work in our world, even as it has changed forms and grown more deceptive. We'll start with its sacred origins.

The Danger of Purity

The Torah contains an account of a strange ritual in ancient Israel. Once a year, on the feast of Yom Kippur, or the Day of Atonement, two male goats were brought to the Temple in Jerusalem. Lots were drawn to determine which goat would be sacrificed to God and which one would be sent away to Azazel, an evil spirit or demon believed to reside in remote regions of the desert.

The high priest would lay his hands on the head of the goat that was bound for Azazel. As he did so, the priest would confess all the sins of the Israelites, symbolically transferring them onto the animal.[6] After the priest had said the appropriate prayers, the people would drive the goat out into the desert, to Azazel, expelling their sins along with it. This goat came to be called, in English, the *scapegoat.*[7]

The idea of the scapegoat was not unique to the Jews, though. The ancient Greeks had their own version of a scapegoating ritual—but they sacrificed humans, not animals. During plagues and other calamities, the Greeks would select a *pharmakós*, a person at the margins of society—usually a cast-off, criminal, slave, or someone thought to be excessively ugly or deformed.

The word *pharmakós* is related to the English word "pharmacy." In ancient Greece, the *pharmakós* was someone initially seen as a poison to the community. The people believed that they had to destroy or expel this person to protect themselves. The elimination of the *pharmakós* was the remedy to the problem. In this sense, the *pharmakós* was both the poison and the cure.

The people often tortured and humiliated the *pharmakós* in a public place.[8] By way of the ritual, they experienced what Aristotle called *catharsis*: the process of releasing strong emotions or impulses through participation in some external event. Aristotle thought catharsis was the purpose of tragic drama. Through it, audience members could release some of their sorrow and pain, thus giving those emotions a safe outlet.

An executive at the investment bank where I once worked organized a paintball excursion in the hills outside our main offices in Hong Kong. "Ah, that was cathartic," Steve said, smiling, when we arrived back at the office. The paintball was not about paintball. Steve knew that if we got to run around and shoot each other with pellets of paint for a couple of hours, we'd be less likely to shoot barbs and insults at each other around the office. Every company needs its own form of cathartic rituals—something more effective than drunken holiday parties. But few companies today are as open about their need for catharsis as the Greeks were.

For the ancient Greeks, the *pharmakós* acted as a substitute, or stand-in, for what they wanted to do to each other. Sometimes the spectacle of humiliating the *pharmakós* lasted for days. The people needed time to release their tensions.

After the ritual was complete, they would unanimously participate in some form of expulsion or killing. In the Greek city of Massalia—today, Marseilles—crowds forced the *pharmakós* to the edge of a high cliff and gathered around him, blocking all routes of escape. They eventually forced him over the edge to certain death.[9]

Because eliminating the *pharmakós* was a collective and anonymous process, the benefits flowed to everyone. Who was responsible for the murder? Everyone, and no one. No single person would feel responsible, absolving each of them from guilt; at the same time, the entire group reaped the benefit of discharging violence onto someone without the threat of retaliation.

Unanimous violence is always anonymous violence. In firing squads, one person's gun is sometimes loaded with a blank so that no one will know whether they fired the killing shot—and so no one exclusively carries the guilt.[10]

There's psychological safety in mobs, just as there are in firing squads. "Can't be sure that I'm responsible" is always a good defense—at least to oneself.

Girard found versions of scapegoating rituals in nearly every ancient culture. The scapegoat is often chosen randomly. But the scapegoat is always *perceived* to be different, marked with some distinguishing feature of an outsider—something to get them noticed.

SCAPEGOAT HOW-TO

CRISIS   SELECTION   DECISION

Scapegoats are often insiders who are perceived to violate the group's orthodoxy or taboos. Their behavior makes them appear as a threat to the group's unity. They come to be seen as cancers or monstrous outsiders who have violated or destroyed the social bonds that hold the group together. Eliminating the scapegoat is the act through which the group becomes unified again.

Nobody is safe from being made into a scapegoat. During a mimetic crisis, perception is distorted. In Freshmanistan, where differences are minor, even the smallest differences are amplified. People project their worst fears onto a scapegoat rather than face the crisis head-on. Nobody wants to pay the price.

Saving People from Themselves

If you're in the ocean when lightning strikes, you have little to fear. But if you're in a pool and lightning strikes it, you have a lot to fear. Celebristan is like the ocean. Freshmanistan is like a pool.

Let's say a huge appliance plugged into an electrical outlet gets thrown off a yacht anchored near a crowded beach in Japan. The high-voltage outlet sends thousands of volts of electricity directly into the water. The electricity would be completely harmless to people swimming off California beaches, thousands of miles away. Most people swimming near the beach in Japan wouldn't feel a thing, either.[11] Water is an excellent conductor of electricity, but electricity is diffused quickly in a large body of water such as the Pacific Ocean. In Celebristan, remember, the degree of social distance between people is high; the risk of mimetic contagion is low.

Now let's say that same appliance falls into a twenty-by-forty-foot pool with twenty people in the water. What might happen? It's safe to say that the consequences won't be the same as they were in the ocean.

The following story will show why mimetic contagion is so dangerous in Freshmanistan. The scene is fantastical and the science is not meant to be completely accurate. I'm using an allegory because it can be adapted to reality in a myriad of ways. The pool is Freshmanistan; the people in the pool are caught in a mimetic crisis; the electricity represents the acute and serious danger that the group has brought on itself—the mimetic contagion that travels quickly from person to person and incapacitates them, making them incapable of solving the crisis on their own.[12]

The logic of how these events unfold is the key—not the logic of reason, but the logic of mimesis.

Here's the scene. Twenty college students, many drunk or well on their way, are playing a spirited game of water polo. One person breaks the rules of play ever so slightly; another retaliates with a harder-than-necessary shove. A fight breaks out between the two of them. Shouting, name-calling, and punches follow. People quickly take sides.

During the scuffle, one of the guys wrestling near the edge of the pool accidentally hooks his arm around the cord of an appliance inexplicably placed only a few feet from the edge. He unknowingly drags the appliance into the pool.

I, like you, am observing this scene at a safe distance, sober enough to see what is about to happen.

Unlike the people on the beach in Japan, the kids in the pool are in serious trouble. Within seconds, the electric current will travel through the water and find all of them. Each will act as a conductor of the electricity for the rest. This is what happens in a mimetic mob.

The English word "contagion" comes from the Latin word *contāgiō*, meaning "with touch." In a mob, contagion happens imperceptibly. As during the community spread of an infectious disease, nobody in a crowd knows who is a superspreader. It's impossible to identify the exact moment when the invisible enemy infiltrates a person's defenses. In the case of mimesis, nobody suspects that their desires are being infected.

We can't predict when the wisdom of the crowd becomes the violence of the mob. We don't see the violent interactions that took place on the other side of the park or the room. We're only a small part of a large system, and nobody on the inside can grasp the dynamics of the whole. What happens in a mob happens in a fog.

Ta-Nehisi Coates describes the fog in a November 2019 opinion piece in the *New York Times*:

> The new cancel culture is the product of a generation born into a world without obscuring myth, where the great abuses, once only hinted at, suspected or uttered on street corners, are now tweeted out in full color. Nothing is sacred anymore, and more important, nothing is legitimate—least of all those institutions charged with dispensing justice. And so, justice is seized by the crowd. This is suboptimal. The choice now would seem to be between building egalitarian institutions capable of withstanding public scrutiny, or further retreat into a dissembling fog.[13]

Our drunken pool revelers, once united by their boisterous play and then by their fighting, are now united in terror. An electric current makes its way through the warm bodies that stand in the shallows. They have five to ten seconds to get out before it's too late, but they are unable to do anything on their own. They are paralyzed by fear and indecision. Nothing has yet catalyzed them to take collective action.

Then an unexpected savior arrives on the scene. One of the guys who had been in the pool earlier, before the fight broke out, has returned from a beer run. He has no idea what just happened.

With a cold beer in each hand, he's smiling and standing near the spot where the appliance is floating in the water. It *zizzes* and sparks as electricity flows from it. He hasn't even noticed.

The people in the pool see their friend standing on the edge of the pool. He's calm. Smiling. They're in danger of death.

One of the guys in the pool points an accusatory finger at him. "He did it!"

The guy holding the beer has no idea what's going on or what he's being accused of. But the eyes of everyone in the pool now fall upon him.

A second finger points and a second voice cries out, "It was him!" Then a third: "He's trying to kill us!"

The fourth and fifth accusations follow quickly.

Accusations are dangerously mimetic.

The first accusation is the hardest. Why? Because there's no model for it. Only in the light of overwhelming evidence would most of us accuse a person of something truly terrible. But in a situation of extreme fear or confusion, the standards change. A person can take on the appearance of an evil perpetrator more easily in a war zone than in a well-run classroom.

The first accusation, even if it's completely false, changes the perception of reality. It affects one's memory and perception of new events. And with each new accusation, there are *more models*. The number of models is why the second accusation is easier than the first to make, the third accusation is easier than the second, the fourth easier than the third.

Models can distort reality, as we saw with Steve Jobs. A mimetic wave of accusation, in which enough people model belief in another person's guilt, can transfigure an accused person before our eyes. We don't see them as they are because they are a mirror of our own violence. In our story, in a single moment, the guy standing on the outside of the pool takes on the appearance of a monster—a murderer—to those in the pool. All because he happened to be in the wrong place at the wrong time.

René Girard recounts the story of the "horrible miracle" of Apollonius of Tyana to show the transfiguring effect of mimesis.[14] Apollonius was a well-known medicine man in second-century Ephesus whose story was recorded by the Greek writer Philostratus. When the Ephesians couldn't end an epidemic ravaging their community, they turned to the great Apollonius. He assured them: "Take courage, for I will today put a stop to the course of the disease." He led them to a theater where there was a blind old beggar who lived in squalid conditions. "Pick up as many stones as you can and hurl them at this enemy of the gods," he said.[15] Apollonius was prescribing the scapegoat mechanism. It may seem odd that he was doing this to put an end to what we think of as something biological, but it becomes clearer if we recognize that it was a mimetic crisis.

In much ancient literature, the line between biological epidemics and psychological epidemics is fuzzy. Girard thought that stories of physical disasters, like plagues, were probably mythologized versions of what really happened: a social crisis, fractured relationships, mimetic contagion.

He found this phenomenon in Fyodor Dostoevsky's *Crime and Punishment*, in which the protagonist, Raskolnikov, "dreams of a worldwide plague that affects people's relationship with each other," as Girard describes it. "No specifically medical symptoms are mentioned. It is human interaction that breaks down, and the entire society gradually collapses."[16] The story of Apollonius and the Ephesians, like many ancient stories, hid mimetic violence—violence that originated in their own community— behind fantastic tales of vengeful gods and demons.

So Apollonius, the medicine man, prescribed his medicine: he told the Ephesians to stone a blind beggar to rid themselves of their disease. Initially, they were shocked by Apollonius's directive. Why would this great healer ask them to kill an innocent man? But Apollonius continued to coax them. Nobody would act. Finally, someone picked up the first stone and threw it.

"As soon as some of them began to take shots and hit him with their stones," Philostratus writes, "the beggar who had seemed to blink and be blind, gave them all a sudden glance and showed that his eyes were full of fire." The Ephesians now saw him as a demon.

After they had stoned the man to death, they found a wild beast in place of his body under their pile of rocks. This symbolizes the transformation that he had undergone in the minds of the crowd.

Peace returned to the city. The Ephesians erected an altar to a god over the place where the event had taken place. The poison had become the cure. Apollonius had taken the Ephesians to the pharmacy. He had provided a *pharmakós*.

Back in the pool, the guy who first spotted and accused the suspect is now filled with rage. He musters all his strength and overcomes the paralysis in his muscles, making his way to the edge of the pool and out of the water.

The electricity represents the dangerous mimetic contagion that binds the people in the pool together and keeps them stuck where they are—unless something or someone comes along that activates an even stronger force. The scapegoat mechanism is the force that breaks it.

The first person out of the pool is a model to the rest. When he acts, the others now have the impetus and motivation to act. Not only have they been activated to flee the deadly water, but now they have friends to pull them out.

(Models spur people to action. Sometimes they help lead to breakthrough performances. At the 2012 London Summer Olympics, more than thirty world records were broken at a time when many scientists thought the human body had reached the limits of its development. In 2019, Eliud Kipchoge ran a marathon in under two hours, something many had predicted wouldn't happen for at least twenty years; now, thanks in part to mimetic desire, we can expect to see that barrier broken again and again.)

As their bodies recover from shock, those who have made their way out of the pool grow more enraged. They converge on the guy who they now believe tried to electrocute them.

The more he protests and reacts to their anger, the more it fuels their rage. "What did I do?" he cries. "I just—"

"Don't lie to us!" they scream.

Everyone is in agreement: the beer guy is the one who knocked the appliance into the pool. Who else could have possibly done it? There was nobody else outside the pool.

And that's their fundamental blindness. They are looking *outside* the pool for answers even though it was their own fighting *inside* the pool that caused the appliance to wind up in the water.

The scapegoat continues to try to explain himself, but the mob hears everything he says as proof they have found their culprit. The louder he protests, the more their anger grows.

A mob is a hyper-mimetic organism in which individual members can easily lose personal agency. Mimetic contagion destroys the distinctions between people—especially the differences in their desires. You can show up to a rally wanting one thing and leave wanting something else entirely.

Crowd psychology is different than individual psychology.[17] Think of people dancing at a rave, moving hypnotically under the sway of the DJ or the band. Author Elias Canetti, who fled Nazi Germany in the late 1930s, wrote about this phenomenon in his masterpiece, *Crowds and Power*, first published in 1960. "As soon as a man has surrendered himself to the crowd," Canetti writes, "he ceases to fear its touch. Ideally, all are equal there; no distinctions count. . . . Suddenly it is as though everything were happening in one and the same body."[18]

None of the kids who were fighting in the pool have ever thought of themselves as violent. But now, in their intoxication and anger, they are ready to inflict serious violence on the beer-runner.

As he begins to realize the predicament that he's in, our unlucky scapegoat begins to look for an escape route. But the crowd is closing in on him.

There are at least three conceivable endings to this story.

In the first ending, the beer runner is expelled from the community. He can no longer show his face in public without being shamed. He's forced to move someplace where nobody has heard the story.

The second ending is the Romantic Enlightenment idea that is not a realistic ending at all but one people still like to imagine. At the peak of their rage, all the drunken pool partygoers recover their faculties of reason, sit down together, and draw up a social contract. They realize that it's plausible that the beer-runner accidentally knocked the appliance into the water. They prohibit him from drinking at future gatherings and mandate that he pay everyone's medical bills.

In the third ending, one of the boys lashes out and strikes the scapegoat in the face, knocking him to the ground. A second guy joins in the beating. And a third and fourth and fifth. Their violence, like all violence, seems righteous. They didn't start it; they're administering justice. In the climax, they do what had become inevitable: they pick up the victim's battered body and throw him into the electrified pool.

In all three scenarios, it is the outsider—literally, the only guy who was outside the pool—who pays the consequences.

In our story, the mob chooses option three: they seize what they believe is justice.

The scapegoat mechanism is most operative at times of instability. Prior to the rise of the Nazi Party, Germany had been thrown into economic and social chaos after its loss in World War I. Other genocides—including but not limited to the Armenian, Rwandan, and Syrian genocides—also came at times of great social instability.

Less conspicuous examples include single, localized incidents of scapegoating in which one person, often someone universally viewed as evil, provides a cathartic relief through his or her death or expulsion. Anthropologist Mark Anspach, in his book *Vengeance in Reverse: The Tangled Loops of Violence, Myth, and Madness*, tells the story of a man who was surrounded by soldiers, stripped of his clothes, and mocked and tormented. His bloody body was dragged through the streets and spat on. Who do you think it was?

After the man's death, which was captured on cellphones and celebrated, the new transitional leader of Libya proclaimed: "All the evils have vanished from this beloved country. It's time to start a new Libya, a united Libya, one people, one future."

The lynched leader was Muammar Qaddafi.

Almost everyone was united in the belief that Qaddafi was a bad man who did evil things. He was. But he could not have been the only wrongdoer in the country. For its transitional leader to claim that all evil had vanished is to make a scapegoat out of Qaddafi. Anspach notes: "The guiltier he is, the more convincingly he can stand in for all other guilty parties and be sacrificed in their stead."

The scapegoating mechanism does not hinge on the guilt or innocence of the scapegoat. It hinges on the ability of a community to use a scapegoat to accomplish their desired outcome: unification, healing, purgation, expiation. The scapegoat serves a religious function.

The Path of Least Resistance

Throughout history, scapegoats have shared some common features. They are people who, for one reason or another, stand out from the crowd and can easily be singled out. In our allegorical pool party, the guy who went to fetch beer unknowingly put himself at risk of being seen as a scapegoat because he was the only one standing outside the pool.

In real life, scapegoats are usually singled out due to some combination of the following: they have extreme personalities or neurodiversity (such as autism) or physical abnormalities that make them noticeable; they're on the margins of society in terms of status or markets (they are outside the system, like the Amish or people who have chosen to live off the grid); they're considered deviants in some way (their behavior falls outside societal norms, whether related to lifestyle, sexuality, or style of communication); they're unable to fight back (this applies even to rulers or kings—when it is all against one, even the most powerful person is impotent); or they appear as if by magic without society knowing where they came from or how they got there, which makes them easy to blame as the cause of social unrest (climate change activist Greta Thunberg's arrival in New York to speak at the United Nations on a zero-carbon yacht marks her as a potential scapegoat).

All scapegoats have the power to unite people and defuse mimetic conflict. A scapegoat doesn't have traditional power; a scapegoat has unifying power. A prisoner on death row possesses power that not even the state governor has. For a family or community in crisis, it can seem like only the death of that prisoner will bring them the kind of healing they seek. The prisoner, then, possesses a quasi-supernatural quality that no one else can stand in for. Only he can heal.

Another distinguishing feature of scapegoats, which Girard named in his 1972 book, *Violence and the Sacred*, is that they are disproportionately kings or beggars—and often both at the same time. If a beggar was chosen to be a scapegoat, he took on a demigod-like quality prior to and after his death because he was seen as the instrument of peace. He had the power to bring about an outcome the people could not bring about themselves. This is why the people of Ephesus built an altar over the spot at which Apollonius had made them stone the blind beggar. Something sacred had happened there.

In *Violence and the Sacred*, Girard explains how *Oedipus Rex* is the story of a scapegoat king.[19] Oedipus was the king of Thebes when the city was hit with a terrible plague. But what kind of plague? What really happened in Thebes? Was there really an outbreak of disease?

According to Girard, we shouldn't put so much trust into the surface-level details of these stories. We have to look deeper. In his view, it's more likely that the city was caught up in a mimetic crisis—"a thousand individual conflicts," he wrote.[20] There may have been a physical plague. Or the social crisis might've *been* the "plague."

Oedipus looks for the murderer of Laius, the previous king (and his father), thinking that he can put an end to the plague if he solves the crime.

To his horror, he learns that he himself killed his father and married his mother. That had to be what caused the calamity as far as the Thebans were concerned. But isn't that odd? Oedipus's crime was social, not biological. He committed patricide and a major taboo by marrying his mother. How could this bring on a plague? This suggests that we have to look deeper, beyond the material explanation.

The story ends with Oedipus gouging out his own eyes and going into exile with his daughter.

We can rightly wonder whether Oedipus is guilty of the crimes of which he was accused. But did he bring on a microbial disease? Of course not. However, in Girard's view, this is the kind of revisionist history that accompanies a scapegoat.

It would be wrong to assume that only people who lived long ago made up these kinds of substitution stories. Have you ever noticed how often people today describe the fallout of crises by employing the language of natural disasters? In 2008, Americans were hit by an *avalanche* of housing debt.[21] Bill Ackman, a prominent hedge fund manager who spoke about COVID-19 on CNBC, said he felt a *tsunami* was coming months before the public started taking the pandemic seriously.[22] According to a White House fact sheet issued on February 4, 2020, migrants were implied to be *flooding* into the country before the president's actions ("President Trump has taken action to end catch and release and stop the surge of migrants flooding to our border," it read). And since the 2008 financial crisis, we talk about companies getting *bailouts* when we refer to the financial lifeline the U.S. government extends to banks and other companies in crisis. To "bail out" is a maritime term that means to remove water with a bucket from a flooded and sinking vessel, usually after an unpredictable storm or other event.

Crises always seem to sneak up on and shock people. With all our modern technology and intelligence, we can't predict them or prevent them. We keep running into crises of our own making.

That's because few people realize when they are caught up in a mimetic process. Most people maintain the illusion of independent desire—the Romantic Lie. But as the world's financial and technological systems become more complex, so do our systems of desire.

Each of us occupies multiple, often overlapping and intersecting, systems of desire. Developing the ability to know which ones we're in and what to do about them is a key goal of the second half of this book.

Mimetic systems are at least as important as physical systems. Some people wonder whether a single butterfly flapping its wings in Japan can cause a hurricane to hit the coast of Florida (the butterfly effect in chaos theory); others wonder whether someone in Russia can create anarchy in the United States with a single Facebook post. The first is a physical system; the second is a system of desire.

What follows is a very short story about a system of desire that nobody understood, and so everyone misidentified. People use mythological language to describe something when they have no idea how it works. That's what myths are for. We need stories to explain the unexplainable. When people find themselves amid disorder with no idea how it got started or of the role that they play in it, they will blame anything—even spiders.

The Dancing Mania of 1518

In July 1518, in the small French town of Strasbourg, a young woman started dancing uncontrollably in the street. John Waller recounts the scene in his book *The Dancing Plague: The Strange, True Story of an Extraordinary Illness*, on which I rely for this story. Her dancing continued for days. The townspeople gathered around her. "They watched as Frau Troffea's dance went on deep into the third day, her shoes now soaked with blood, sweat trickling down her careworn face," writes Waller.[23] Within days, more than thirty people had taken to the streets with the same uncontrollable urge to dance. The chief magistrate, the bishop, and the doctors forcibly admitted some of them to the hospital. But the cause of and cure for the erratic dancing remained obscure.

Theories about what caused the Dancing Plague of 1518 circulated for decades. Mental derangement and demonic possession were two popular ones. But neither explained why the behavior seemed to spread from the original dancer to those around her. What could account for the social contagion?

Spontaneous dancers popped up in cities throughout Europe. In some places, people started dancing only at specific times of the year. In the Puglia region of southern Italy, every summer brought outbreaks of hysterical dancers, which the Italians called *tarantati*. Most believed the dancing was the symptom of a disease caught through the bite of a tarantula, or *taran-tola*, which made a person imitate the spider's movements.

Strange rituals developed, including the belief that dancing to a specific song and following a specific rite, a kind of liturgy, was the only way to cure the disease. People would gather around the infected person in a room or town square, play music, and cheer the person on as they attempted to dance to the spider's rhythm. If they danced in a way that satiated the creature, it was believed, its influence would end.

An infected person, a *tarantata*, took on a quasi-sacred quality. Men and women afflicted with tarantism were pariahs—the cause of calamity and fear in the community—but they were also the only ones who had the power to restore order.

But was any of it true?

For centuries, nobody could put their finger on the real cause of the dancing. It wasn't until the Italian cultural anthropologist and ethno-psychiatrist Ernesto de Martino went to Puglia in the 1950s that a truer picture began to emerge.[24] He learned from his interviews with hundreds of locals that people who danced had something in common: most had been through some kind of trauma. The trigger for the dancing seemed to be a crisis—an impossible love, a forced wedding, a job loss, the transition into adolescence, or something else that upset the dynamic in their lives and, through mimesis, in the community.

The affliction seemed to be *relational*. De Martino's work exposed the hidden power relations, social tensions, and unacknowledged crises of desire.

Tarantism was a religious ritual that had the effect of restoring order from social chaos. The spiders were scapegoated. The rituals that expunged the spider's influence from the afflicted dancers brought everyone together for a cathartic experience that supposedly purged a disease from their midst. As strange as the ritual was, it functioned to protect the community from an even bigger social crisis—the further breakdown of relationships. Think of the manic dancing as an alarm bell—maybe even a church bell—that signaled it was time for everyone to come together and exorcise their demons.

While tarantism gradually died out, its cultural vestiges remain. The popular folk dance called the tarantella, one of the most popular forms of folk music in southern Italy today, comes directly from the dance used in the five-hundred-year-old rituals.

Why did the people in southern Italy lie to themselves about the real cause of their crises? Claiming that people had the "spontaneous" urge to dance after a spider bite is a version of the age-old Romantic Lie. The dancing was caused by mimetic desire. It was symptomatic of a social contagion. A scapegoat was needed to restore order. In this case, it was a tarantula.

Tarantulas, like most wolf spiders, will only bite a human if continually provoked, by the way. Their venom causes mild swelling, light pain, and itching.

Safety in Judgment

Scapegoats are chosen through a mimetic process of judgment, not a rational one.

Consider the ancient practice of stoning: a group of people throws stones at someone until they die from blunt trauma. It was the official form of capital punishment in ancient Israel—the Torah and Talmud codify it as punishment for certain offenses—but its origins are even older.

The practice of stoning, in its most primitive form, happened spontaneously. It occurred outside of what we now know as "due process." (Our modern conception of due process—that a person not be stripped of any freedoms or punished prior to a due process of the law—originated in written form in England's Magna Carta, in 1215.)

The phrase "casting the first stone" is known by nearly everyone in the Western world. What is it about the first stone that matters so much?

The phrase comes from a rabbi in first-century Palestine, Jesus of Nazareth, who was present at the strangest stoning in world history—strange because the stoning *never happened*, yet we know more about it than most other stonings. That we know about this two-thousand-year-old non-stoning at all is remarkable. What makes it so important? It's a story about mimesis and the scapegoat mechanism.

Jesus came upon a woman who had been caught in the act of adultery and who was about to be stoned by an angry mob. He intervened, saying, "Let anyone among you who is without sin be the first to throw a stone at her."

The words threw everything off balance. The cycle of destructive violence was knocked off its course. One by one, the men standing around the woman began dropping their stones and walking away. First one, then another, then the pace accelerated.

What happened? Why was throwing the first stone so hard? Because the first stone is *the only stone without a mimetic model.* The thrower of the first stone, often acting in a violent rage, gives the crowd a dangerous model to follow. As we saw earlier in the story of Apollonius and the Ephesians, once the first stone is thrown, the second stone becomes easier to throw. It is always easier to desire something—even, and maybe even especially, violence—when it has been desired by someone else first.

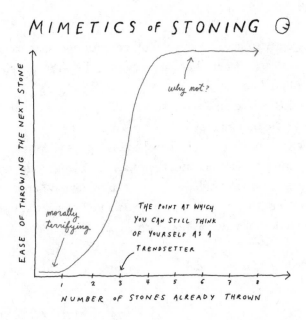

MIMETICS of STONING

why not?

THE POINT AT WHICH YOU CAN STILL THINK OF YOURSELF AS A TRENDSETTER

morally terrifying

EASE OF THROWING THE NEXT STONE

NUMBER OF STONES ALREADY THROWN

The first stone thrower shows the way. The second reinforces the desire. Now the third person in the crowd is hit with the mimetic force of *two* mimetic models. They cast the third stone and become the third model. The fourth, fifth, and sixth stones are cast with relative ease compared to the first three. The seventh is effortless. Mimetic contagion has taken hold. The stone throwers become unattached from any form of objective judgment because their desire for a scapegoat has overpowered their desire for truth.

Anger metastasizes and spreads easily. In a study conducted in 2013 and published in 2014, researchers at the University of Beijing analyzed influence and contagion on Weibo, a popular social media app in China. They

Tactic 7

ARRIVE AT JUDGMENTS IN ANTI-MIMETIC WAYS

If you're taking a poll or a vote in a public place, it's essential that people cannot see how other people are voting—*if* you want anything resembling a true, pre-mimetic reflection of what people think, that is. The mimetic influences are too strong. It's important to find ways to allow each member of a group to arrive at a verdict—whether it be an investment decision or that of a courtroom jury—through the most independent process possible.

found that anger spreads faster than other emotions, such as joy, because anger spreads easily when there are weak ties between people—as there often are online.[25]

Many people have died as a result of road rage. Nobody to my knowledge has ever died of road joy.

The tactic Jesus used to prevent the stoning was depriving the crowd of a violent model and replacing it with a nonviolent model. Instead of a violent contagion taking hold, a nonviolent contagion happened instead. The first person dropped their stone. Then, one by one, the rest followed. Cycle 1, mimetic violence, was transformed into Cycle 2, a positive mimetic process.

Both depended on a model.

The Joy of Hate Watching

For more than twelve years, tens of millions of Americans watched the same TV show. Battle lines were drawn at the beginning of each episode. Every person on the show wants the same thing: the prestige of being proclaimed the victor, which will earn them praise from an authority figure and, with it, the adulation of the masses. And each of them is willing to do nearly anything to get it.

They fail. They engage in finger-pointing, backstabbing, and betrayals. Then, when the game is over, they walk into a giant boardroom. Donald Trump is seated, scowling, at the middle of a long table. They all want to be his next apprentice, but only one can win.

Trump lets the mimetic crisis escalate until it's boiling over. Finally, he points a finger at one of them and says, "You're fired!" The crisis is averted. The scapegoat goes home. The team can get back to business.

Meanwhile, the perception of Trump as a mimetic model—a person who *knows what he wants*—grows stronger every time he points his finger and utters the words "You're fired."

After a dozen years of Trump cultivating and cementing his status as the "master" and everyone else as the "apprentice," it's not surprising that he became a cult-like figure. He single-handedly resolved a mimetic crisis by bringing order during each of the 192 episodes of *The Apprentice* (including *The Celebrity Apprentice*, which received even higher ratings). As we'll see shortly, there is probably no more effective thing that a politician—or potential politician—can do to gain popular support than resolving a mimetic crisis. Such a person mimics the role of the High Priest in ancient Israel.

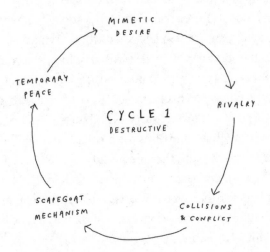

According to Girard, the scapegoat mechanism happened spontaneously in ancient societies. Eventually, these societies began ritually reenacting the process that led to the scapegoat mechanism—creating disorder, allowing mimetic tension to reach a peak, then expelling or sacrificing something symbolic. (This is the formula of reality television today.) They found that catharsis flowed to everyone.

These rituals worked due to sacrificial *substitution*. Humans realized they could substitute an animal for a human. The sacrifice of animals has gradually been replaced by the termination of executives, mass incarceration, and social media cancellations. There seems to be no limit to human ingenuity when it comes to satiating our hunger for sacrifice.

Substitute sacrifices permeate our culture. They have seeped into sports, organizational life, universities, and literature.

Stephen King's first novel, *Carrie*, is a terrifying account of the scapegoat mechanism gone wrong. At the end of the novel, a bullied high school girl named Carrie takes revenge on her classmates, causing death and carnage with her telekinetic powers after she is humiliated on prom night.

In conceiving the novel, King reflected on the type of girl he had in mind to play the protagonist. "There is a goat in every class, the kid who is always left without a chair in musical chairs, the one who winds up wearing the KICK ME HARD sign, the one who stands at the end of the pecking order," King remembers. He used "the two loneliest, most reviled girls" in his own high school class as models for Carrie—"how they looked, how they acted, how they were treated." One of them later died while having a seizure. The other shot herself in the abdomen after the birth of a child, according to King.[26]

King's genius was turning a would-be scapegoat into a horrifying figure who had the power to enact revenge. Real-life scapegoats do not.

Shirley Jackson's 1948 short story "The Lottery" is about a community that convenes annually to draw lots for a ritual stoning. The scapegoat ritual is designed to make sure that the bounty of the harvest continues—in other words, to maintain peace. Ritual sacrifices do not literally bring divine blessing on a harvest; they can, however, resolve mimetic tension between humans competing for scarce resources. A similar dynamic is at work in the 2019 horror film *Midsommar*.

William Golding's 1954 novel *Lord of the Flies* depicts a mimetic crisis among adolescents trapped on a remote island. One of the kids, Piggy, is consistently made to bear the punishment for the entire group's sins and suffers in their place.

The plot of the *Hunger Games* films revolves around a sporting spectacle in which boys and girls ages twelve through eighteen are chosen by leaders in the capital of a dystopian country, Panem, to compete to the death. The games concentrate all of the society's internal conflicts onto a select few who are forced to bear out the violence on behalf of the rest.[27]

I don't know if any of these writers were explicitly thinking about the scapegoat mechanism when they wrote their stories. But the pervasiveness of the theme is striking, and it's worth asking whether it points to an underlying truth. Girard thought that truth was scapegoats.

Professional sports thrives on them to engage fans. American football games are a sacred rite in which two teams and their fan bases reenact a crisis of undifferentiation. The league is structured to promote parity; on any given Sunday, any team can win. Hours of pregame analysis followed by the opening kickoff bring the tension to its height. The entire season plays out like this in a series of dramatic ups and downs. After the last game, losing teams' coaches are fired, players are not re-signed. ESPN covers the drama and dysfunction within organizations. Once someone has been purged, ridding a team of its plague, the organization can move on with a fresh shot at glory.

Terrell Owens, a star receiver in the NFL, became a scapegoat for every team he played on later in his career. In basketball, David Fizdale, fired by the New York Knicks in 2019, was one of a long list of Knicks coaches drummed out as a scapegoat over the past two decades (becoming the head coach of any New York pro sports team is tantamount to signing up to be an eventual scapegoat). Even the legendary Phil Jackson, who coached Michael Jordan and Kobe Bryant, left as a scapegoat. And who can forget Steve Bartman? During the 2003 National League Championship Series, the Cubs fan interfered with a player on his own team who was trying to make a catch near the stands, resulting in a missed catch at a critical moment in the game. Bartman had to go into hiding, and the physical baseball involved in the incident was publicly detonated by a special effects expert in 2004; the remains were boiled down and used in a pasta sauce in 2005.

Are scapegoats the problem? Or are scapegoats the solution?

C. P. Cavafy's poem "Waiting for the Barbarians" is about a community in crisis. They gather in the main square, brought together by the news of an impending attack by neighboring barbarians. The barbarians never show up. "Now what's going to happen to us without barbarians?" reads the penultimate line of the poem. It closes: "Those people were a kind of solution."

The Scapegoat Wins

Caiaphas, a first-century Jewish high priest, was the greatest politician in history according to René Girard. Not necessarily in a noble sense, but in the sense that he knew exactly what needed to be done to satisfy all of the stakeholders in any dispute and quell social unrest. According to Girard, Caiaphas was simply putting into political practice what the scapegoat mechanism had always done: he was turning to it "as a last resort to avoid an even greater violence."[28]

When Jesus was arrested in Jerusalem, Caiaphas held a secret meeting with the chief priests and the religious and political councils. They had to figure out what to do about the man from Nazareth. He was creating tension in Jerusalem at an already tense time. Rifts were opening up in all parts of the society. There were dozens of splinter groups and sects forming. Jesus came out of nowhere, from a backwater town, and lived at the margins of society, breaking cultural norms and challenging the power of authorities. So the challenge facing Caiaphas concerned not only Jesus. The challenge was how to preserve Israel.

As the meeting convened, Caiaphas sat back and watched the others go back and forth, proposing hypothetical questions and abstract ideas lacking in substance and concreteness, indecisive. Finally, he'd had enough.

"You know nothing at all!" he cried. "You do not understand that it is better for you to have one man die for the people than to have the whole nation destroyed."[29]

Caiaphas could not have known the full import of what he was saying. "A scapegoat remains effective as long as we believe in its guilt," wrote Girard in his final book, *Battling to the End: Conversations with Benoît Chantre*. "Having a scapegoat means not knowing that we have one."[30]

So Caiaphas could not have thought of his plan as the prescription of the scapegoat mechanism. Yet he must have known that targeted violence against a potent symbol is effective at quelling a turbulent crowd. Killing Jesus would satiate people; it would unify them and prevent the crisis from escalating.

Caiaphas won support for his idea. And within days, Jesus was crucified.

In making his recommendation, Caiaphas was being utterly practical. He was recommending a ritual sacrifice (crucifixion) to achieve a specific outcome (increased unity and peace). That is not unexpected of a religious leader. "The goal of religious thinking is exactly the same as that of techno-logical research—namely, practical action," wrote René Girard in *Violence and the Sacred*.[31] He viewed the scapegoat mechanism as the epitome of a religious or sacred act.

When Girard writes that "religious thinking" has the goal of practical action, he is not disparaging religious belief in any way—he is referring to the *sacrificial mentality* that people bring to problem solving. Nearly all people are religious in the sense that they subconsciously believe that sacrifice brings peace.

Consider how ingrained sacrificial thinking is in our psyche. If only we could destroy that other political party, that other company, those terror-ists, that troublemaker, that fast-food joint next door that has caused me to gain ten pounds, everything would be better.[32] The sacrifice always seems right and proper. Our violence is good violence; the violence of the other side is always bad.

For many years, according to Girard, sacrificial rituals were so effective that they hindered scientific progress. "We didn't stop burning witches be-cause we invented science; we invented science because we stopped burn-ing witches," Girard said in a 2011 CBC interview with David Cayley. "We used to blame droughts on witches; once we stopped blaming witches, we looked for scientific explanations for drought."[33]

Humanity still tends to revert to a primitive, sacrificial mindset that characterized our ancestors and kept them stuck in cycles of violence. From the perspective of the crowd, the scapegoat mechanism is entirely rational. So when the scapegoats become the sacred center around which a culture turns—when myth and superstition reemerge as dominant forces in a culture—actual rationality takes a back seat.

In Girard's view, an understanding of the scapegoat mechanism was de-veloped in the historical unfolding of Judaism and Christianity—both in the clarity that the biblical stories offer into the innocence of scapegots and in how, over the past 2,000 years, the scapegoat mechanism seems to have grown less effective at bringing even illusory peace. The Jewish and Chris-tian scriptures contain very peculiar accounts of scapegoats. The difference between their accounts and others is shocking: in the scriptures, mimetic desire seems to be specifically addressed, and stories of scapegoats always

seem to be told from the perspective of the accused. It is such an overturn-
ing of traditional accounts that even those who are very familiar with these
scriptures don't realize how different they are.[34]

Before we get to the revelation of the scapegoat mechanism, though,
notice the allusion to mimetic desire as far back as the Decalogue, or Ten
Commandments. In the book of Exodus, the Tenth Commandment is
striking. It seems to directly prohibit mimetic desire of any kind:

> You shall not covet your neighbor's house; you shall not covet your neigh-
> bor's wife, or male or female slave, or ox, or donkey, or anything that
> belongs to your neighbor. (Exod. 20:17)

While the rest of the prohibitive commandments forbid *acts*, the tenth
commandment forbids a certain kind of *desire*. Girard notes that the He-
brew word that is often translated as "covet" means something simpler:
"desire." Read through the lens of desire, these biblical stories take on a rich
anthropological meaning.[35]

But if mimetic desire is universal, a constituent part of who we are as
humans, how could one of the Ten Commandments forbid it? The Tenth
Commandment forbids *rivalrous* desires. It prohibits them because they
lead, as we've seen by now, to violence.

The rest of the scriptures read like a playing out of that violence, with
warnings along the way. In the Torah, the most salient example of the
scapegoat mechanism is found in the story of Joseph, son of Jacob. Joseph
is sold into slavery in Egypt by his eleven older brothers, who are envious
that he is Jacob's favorite son. It is a case of all against one. In any other
telling of the story, like at our pool party, Joseph would've been accused of
something that warranted his expulsion and sacrifice. In the biblical narrative, though, everyone can see that he is completely innocent.

Once he arrives as a slave in Egypt, he secures his release from prison,
eventually earns the respect of the country's leaders, and is placed in a
position of power. Then it happens again. Joseph, the foreigner, is falsely
accused of a crime. But again, he is thoroughly vindicated to the reader.
Time and time again, Joseph is shown to be the innocent victim of unjust
accusations and violence.

Finally, at the end of the story, Joseph has become the vizier, second
in power only to the pharaoh. His brothers come to Egypt begging for aid

during a long famine, and they interact directly with Joseph without recognizing him.

Joseph does not want to return violence for violence. He is not like Alexandre Dumas's Count of Monte Cristo, who carefully plots revenge on everyone who ever crossed him. Instead, Joseph forgives his brothers. Not without testing them, though.

Joseph sets up his brother Benjamin as a thief, and Benjamin is arrested. Joseph makes the brothers believe that he will take the innocent Benjamin and do what he wishes to him as punishment for the transgression. But one of the other brothers, Judah, steps in and volunteers to take Benjamin's place. That allows Joseph to see that they have changed. The destructive cycle has been broken. Touched by this act, Joseph reveals his true identity to them. Both Joseph and Judah refuse to participate in the scapegoat mechanism.

Already, then, in the book of Genesis, Girard sees the scapegoat mechanism unmasked. He would see its complete unmasking in one particular event that would come much later in the scriptures.

Girard urged everyone, regardless of their religious beliefs (or lack thereof), to pay attention to what happened at the crucifixion of Jesus. Girard read this story primarily as an anthropologist. What he found was human behavior operating differently than he had seen anywhere else in his reading of history.

The mob attempted to make Jesus their scapegoat. But the mechanism was subverted in a radical way—which is one of the reasons it has had such enduring cultural significance, even from a historical point of view alone.

The crucifixion of Jesus failed to unite a community unanimously against a scapegoat. It did the opposite—it caused enormous division. For a short period of time, the crucifixion *seemed* to have the desired effect. The mob was quelled, and order was temporarily restored. But very shortly after Jesus's death, a small number of people—those who knew Jesus intimately—came forward to proclaim his innocence and said that he was alive.

A division opened up between those who wanted to preserve the old sacrificial order and those who saw the scapegoat mechanism for what it is: an unjust sacrificial mechanism.

The gospel texts are radically different from Greek, Roman, and other common myths. In the pagan accounts of unanimous violence, the reader or listener gets the impression that the violence was done to someone guilty, deserving of punishment. That's because the only people left to tell the story are the scapegoaters. The stories are told from the standpoint of the persecutors, who honestly believed in the guilt of the scapegoat.[36] In the crucifixion of Jesus, the reader is meant to identify with the crowd, but *also to see the folly of the crowd* and to move beyond it—to finally, for the first time, grasp the truth about human violence.

I told the story of the pool party from the standpoint of an omnipresent narrator who knew that the guy who went on the beer run was innocent. If one of the murderers had been the narrator instead of me, you never would've known that the scapegoat had done nothing to merit the anger of the crowd. You would have heard only one interpretation of the event and not even known to look for another. It would've been the same story from every single person who had been in the pool: the victim was guilty.

I told the story as someone who was *aware of the scapegoating mechanism*. This is how the gospels worked. For the first time in history, the story was told from the standpoint of the victim. Girard sees this as a definitive turning point—the moment when the scapegoat mechanism began to lose its absolute power. The story forces people to come to grips with their own violence. A veil was lifted on the recurring cycle of violence in human history.[37]

That lifting of the veil, as we all know, did not put an end to violence. The revelation has worked its way through time slowly. But the revelation is not reversible. If the modern world seems to be going crazy, it's partly because we are hyperaware of the ways in which exploitation and violence against innocent victims occur, but we simply don't know what to do about them. It's like we've been told something terrible that we didn't want to know, and which we're powerless to fix entirely on our own. And that's a recipe for collective madness.

Those of us who grew up in cultures that have been touched by this history have been so inculcated with the concern for innocent victims that it's easy to forget how some of our deeply held convictions may have been formed in the first place.

Some things, once they have been seen, can never be unseen.

Self-Awareness, Self-Hatred

"Examine ancient sources, inquire everywhere, dig up corners of the planet, and you will not find anything anywhere that even remotely resembles our modern concern for victims," wrote René Girard.[38] Think about how peculiar it is.

At the present time we have such a heightened sensitivity to innocent victims that we find new injustices to accuse ourselves of daily. We are made highly uncomfortable by the thought that someone being treated harshly might be innocent. Where did this passionate spirit of defending victims come from?

Did it come merely from the Enlightenment—the conceit that we are now smarter, rational people who can judge the past rightly from our heightened, enlightened perch? Or did it come from something else entirely?

According to Girard, our cultural awareness comes from the biblical stories. The awareness couldn't have come about by thinking about it hard enough. We had a blind spot because we were part of the crime. The events recounted in the Bible showed us something that no amount of reasoning had arrived or could arrive at: the innocence of victims.

We were like a person with a nail stuck in their head, raking our brains wondering what is causing our headache—until someone simply holds a mirror to our face. The effects hold for everyone who grew up in a culture that has been touched by these stories in some way, even if you are not familiar with them yourself, because they have had thousands of years to seep deeply into the fabric of our lives.

Western culture has developed strongly around the defense of victims. Over the past two thousand years, there have been dramatic advances in public and private law, economic policies, and penal legislation to protect the vulnerable. Civilian (not military) hospitals sprang up in the fourth century.[39] Monasteries in the Middle Ages protected the aging and dying, travelers and orphans. They served as what today we might call a social safety net. They protected victims. Today, both the pro-life and pro-choice movements speak the language of victims in their own way. No language is more powerful.

One of the great ironies of the modern world is that Western democracies like the United States, in which there is a separation between church and state, have made the defense of victims an absolute moral imperative even as they have largely expunged religion from public life. It's as if they said, "We'll take your defense of innocent victims, Jews and Christians, and we'll raise you—we will do even better than you have at defending victims." And in many cases they do.

In response, many religious people have taken secular culture as a mimetic rival. The culture wars are a giant mimetic rivalry with many faces, a hydra with a thousand heads—a rivalry from which each side would be wise to extract itself.

The development of human rights as we know it was born partly from the indirect acknowledgment that anyone can become a scapegoat under the right circumstances. After approximately 75 million people were killed in World War II, the United Nations issued the Universal Declaration of Human Rights, which protected fundamental human rights applicable to all people; it was translated into more than five hundred languages and dialects. The creation of the declaration stemmed, in large part, from the appalling number of innocent victims made during the course of the war.

These developments have dramatically shifted the balance of power. Previously, most victims were totally powerless to defend themselves. Today, nobody has more cultural influence than someone who has been recognized as a victim. It's as if the poles of the earth's magnetic fields changed places, the way they do every few hundred thousand years. The scapegoat mechanism has been so thoroughly subverted that there is some semblance of a *reverse scapegoating mechanism*, whereby an innocent victim is recognized as having been treated brutally and then a wave of support swells up around that person.

The original scapegoat mechanism brought order out of chaos—but the order depended on violence. The reverse process brings chaos out of order. The chaos is meant to shake up the "orderly" system, predicated on violence, until something serious is done to change it. The death of George Floyd in the United States in May 2020 is one salient example.

Obviously, the defense of victims is a good thing. At the same time, it brings new dangers. In the same way that scapegoating rituals in archaic religions were entirely practical—that is, they were used to achieve practical ends—so too can the defense of victims be used for practical purposes. James G. Williams, in his foreword to one of Girard's most well-known works, *I See Satan Fall Like Lightning*, attempted to sum up Girard's thinking on this point: "Victimism uses the ideology of concern for victims to gain political or economic or spiritual power," he wrote. "One claims victim status as a way of gaining an advantage or justifying one's behavior."[40] Victims now have the power to make new scapegoats of their own choosing.

An open and honest memory is needed to prevent that power from becoming tyrannical.

The prophets of ancient Israel were systematically ridiculed and scapegoated. Many of them were killed. The Pharisees, a religious sect in first-century Palestine, revered those ancient prophets and built monuments to them. The Pharisees railed against violence and followed the law meticulously. They claimed that if they had been living in the days of their ancestors, they would not have killed the prophets.[41]

And then they collaborated in killing Jesus.

This is the precarious mindset of those alive today who look back at people living in Nazi Germany, or in Soviet Russia, or in 1950s America, or at the time of Christ, and swear that they could never have participated in such ideology or racism or demagoguery. That is exactly what makes the scapegoat mechanism possible—the idea that you are not capable of it. We lack the humility to see that we are all caught up in mimetic processes.

Aleksandr Solzhenitsyn, who spent eight years in Soviet forced labor camps (the Gulag) and who saw his country descend into incoherence and evil, reflected later in life: "If only it were all so simple!" he wrote. "If only there were evil people somewhere insidiously committing evil deeds, and it were necessary only to separate them from the rest of us and destroy them. But the line dividing good and evil cuts through the heart of every human being. And who is willing to destroy a piece of his own heart?"[42]

Signs of Contradiction

As mentioned earlier, one of Jenny Holzer's billboards in Times Square pleaded: "PROTECT ME FROM WHAT I WANT." It drew attention because it was a sign of contradiction. Through its stark contrast with its surroundings, Holzer's art drew communal attention to its message. And through it, people were drawn toward a more honest examination of themselves. The message led not to rivalry and blame and violence but to self-reflection and maybe even transformation. Consumer culture did not have to have the last word.

The crucifixion of Jesus likewise stands at the center of human history in stark contrast to everything that surrounds it: the politics of the Roman Empire, the violent execution of criminals, and the prevailing narrative. It prompts us to do an honest examination of our own role in sustaining a cycle of violence. So do the world's new scapegoats, which are made daily—if we have the eyes to see.

American author Ursula K. Le Guin wrote a short novel in 1973 called *The Ones Who Walk Away from Omelas*. The story takes place in a fictional, utopian city of "happiness" called Omelas. We aren't told where it is, or even what time period we're in. All we know is that all of its citizens have found a way to structure their society to maximize the happiness of all.

All except one, that is.

Midway through describing one of the citizens' summer festivals, the narrator reveals a dark secret: the entire functioning of the city, and all of its happiness, depends on the expulsion, entrapment, isolation, and perpetual misery of a single child kept prisoner underneath the city.

When the citizens of Omelas are old enough to learn the truth about their city, they're shocked and disgusted. In time, though, most come to accept this injustice for the sake of the city's happiness.

And yet a few citizens walk away. The story ends with the narrator describing the destination toward which these few people walk: "The place they go towards is a place even less imaginable to most of us than the city of happiness. I cannot describe it at all. It is possible it does not exist. But they seem to know where they are going, the ones who walk away from Omelas."[43]

The whole town knew about the child under the city, but only a few left. The rest accepted the compromise. Most people do.

"Each person must ask what his relationship is to the scapegoat," wrote René Girard. "I am not aware of my own, and I am persuaded that the same holds true for my readers. We only have legitimate enmities. And yet the entire universe swarms with scapegoats."[44]

Part II

THE
TRANSFORMATION
OF DESIRE

We can now go out into the world and find all of the scapegoats, point out every rivalry, and maybe even have a laugh at the expense of those still caught up in the quixotic throes of mimetic desire. But be careful. The scapegoat mechanism works by diversion; the more we see it in others, the less we can see it in ourselves.

Sure, we can "use" mimetic desire, just as we can use other people's trust or hearts or bodies. It might help us recognize the next Facebook. Or become better pickup artists. Or make a lot of money in a volatile stock market.

But it will be a Faustian bargain. It will ensnare us in rivalries and keep us from doing the hard work of finding and pursuing those desires that ultimately lead to fulfillment.

David Foster Wallace once mused that to live in a world in which the internet encompasses more and more of our lives, with increasingly sophisticated porn (like a virtual reality version), "We're gonna have to develop some real machinery, inside our guts, to help us deal with this."[1]

Machinery that might also help us respond in a better way to the images we see on the twenty-four-hour news cycle, to the polarized political environment, and to other mimetic accelerants such as frictionless technology that remove every barrier to restraint.

We're going to have to develop some machinery in our guts to help us resist dangerous mimesis. That will require being somewhat anti-mimetic, which we'll explore in Part II.

What does this mean? It's not that we should be (or can be) free of mimetic desire. Being anti-mimetic is not like Nassim Nicholas Taleb's "antifragile"—it's not merely the opposite of mimetic. Being anti-mimetic is having the ability, the freedom, to counteract destructive forces of desire. Something mimetic is an accelerant; something anti-mimetic is a decelerant. An anti-mimetic action—or person—is a sign of contradiction to a culture that likes to float downstream.

The second half of this book is about developing the stuff inside our guts. It's about developing the ability to resist knee-jerk social reflexes, to separate from the deafening crowd, to renounce the allure of easy desires to want differently, and to want more.

Chapter 5

ANTI-MIMETIC
Feeding the People, Not the System

Questioning Desire . . . Probing Origins . . . Dropping Out

"What do you fear, lady?" he asked.

"A cage," she said. "To stay behind bars, until use and old age accept them, and all chance of doing great deeds is gone beyond recall or desire."

—J. R. R. Tolkien

"There are key moments in life when we ask ourselves questions," Sébastien Bras says. Bras is a celebrated chef whose flagship restaurant, Le Suquet, draws crowds in spite of its location in the middle of nowhere. "Questions like 'What have we done in the past, where are we today, and what do we want for tomorrow?'"

His office windows provide a 180-degree view into his kitchen, which is humming with staff preparing for that night's dinner service. But I barely notice them. Bras is commanding, and he speaks deliberately. In response to several of my questions, he begins by stating how many points he'll make on the subject—he thinks in lists, like recipes.

He says that he wants to tell me about three key moments in his career: first, when his father, Michel Bras, opened the restaurant in 1992 on the Aubrac plateau in the south of France; second, when Michel first earned three Michelin stars in 1999; and third, the day in 2009 when Sébastien first sat down in the chair he's sitting in now, behind what used to be his father's desk, marking the transition of the restaurant from one generation to the next.

But now there was a fourth important moment. In June 2017, Sébastien told the Michelin Guide—the hallowed, 120-year-old institution that had awarded Le Suquet its highest distinction, three stars, for nineteen straight years—that he was no longer interested in their stars or their opinion. He asked them to remove his restaurant from the guide.

How does someone cease to want something that they wanted their entire life?

Moving Goalposts

Author James Clear writes in his book *Atomic Habits: An Easy and Proven Way to Build Good Habits and Break Bad Ones* that "we don't rise to the level of our goals, we fall to the level of our systems."[1] From the standpoint of desire, our goals are the product of our systems. We can't want something that is outside the system of desire we occupy.

The obsession with goal setting is misguided, even counterproductive. Setting goals isn't bad. But when the focus is on how to set goals rather than how to choose them in the first place, goals can easily turn into instruments of self-flagellation.

Most people aren't fully responsible for choosing their own goals. People pursue the goals that are on offer to them in their system of desire. Goals are often chosen *for* us, by models. And that means the goalposts are always moving.

Some trends in goal setting: don't make goals vague, grandiose, or trivial; make sure they're SMART (specific, measurable, assignable, relevant, and time-based)[2]; make them FAST (another acronym: frequent, ambitious, specific, and transparent)[3]; have good OKRs (objectives and key results)[4]; put them in writing; share them with others for accountability. Goal setting has become very complicated. If someone tried to take all the latest tactics into account, it would be a wonder if they managed to set any goals at all.[5]

Don't get me wrong: some of these tactics may be helpful. If I want to lose weight, it would help to set goals that are specific, measurable, assignable, relevant, and time-based. But it's not immediately apparent that losing weight is a good goal for me to begin with. Why do I want to lose weight? What if I am at an ideal weight and I want to lose weight simply to look more like someone I saw on Instagram?

People set goals and make plans to arrive at a future point called "progress." But will it be progress? How can we be so sure? Sébastien Bras set a goal to maintain his restaurant's three Michelin stars, and he pursued it vigilantly. Then one day he realized the pursuit was killing him. Some goals—even good ones—overstay their welcome.

Have you noticed that goals have an irreproachable and unimpeachable status? You want to run an ultramarathon? People will applaud your determination. Run for city office? You have their support. Sell your home and move into the back of a van? Cool, essentialism is in. Nobody will question your goals.

But it's worth asking where goals come from in the first place. Every goal is embedded within a system.

Mimetic desire is the unwritten, unacknowledged system behind visible goals.[6] The more we bring that system to light, the less likely it is that we'll pick and pursue the wrong goals.

Mimetic Systems

The U.S. education system, the venture capital industry, the publish or perish racket for academics, and social media are examples of mimetic *systems*: mimetic desire sustains them.

In U.S. secondary schools, most students organize their energy around college application builders such as their grade point average, standardized test scores, and extracurricular activities. Many high schools have the goal of 100 percent "college placement," even though many university students feel they are no longer getting value for their money and wind up crushed by debt.

Students have lost sight of the teleology, or final purpose, of the education system.[7] When you're in fifth grade, you know clearly that your goal is to get to sixth grade—and it goes like that up through twelfth grade, at which point you've spent the past four years of your life preparing for something called "college" along carefully defined lines (you probably even had a college advisor who advised you as to which schools you should apply for, based on your data).

College is where the teleology grows even less clear. Is the goal to get a good job? To get into grad school? To be a well-rounded person who is able to think critically? To be a good citizen? When I started at the Stern School of Business as an undergrad, I had no idea. So what did I do? I looked around to see what everyone else was doing—what everyone else seemed to *want*. There was a clear object of desire: Wall Street. So I fought for it, and I got what I thought I wanted. And that's when I began my miserable fifteen-month career in Advanced Excel and PowerPoint.

Traditional venture capital (VC) funds operate in a mimetic system. They need extraordinary returns on their investments to justify the risks they take. Many only fund companies that have the potential to return ten times the value of their investment within five to seven years. Because of their investment timeline, VCs favor technology companies that can scale quickly—not food service companies that might grow steadily but only incrementally over twenty or thirty years. They're looking for instant ramen, not risotto.

The VC demand for quick-hitting investments increases the attractiveness of tech start-ups to entrepreneurs. A mimetic system takes shape. It is driven not only by economic incentives and financial returns—which no doubt factor in—but also the prestige and validation that come with being financed by the right VC. They award Michelin stars in the form of investment checks. And for VCs: the benefits of having invested in sexy companies and headline-grabbing CEOs.

Social media platforms thrive on mimesis. Twitter encourages and measures imitation by showing how many times each post has been retweeted. People are more likely to use Facebook the more they are engaged with mimetic models, rivals whose posts they can track and comment on.

The greater the mimetic forces on a social media platform, the more people want to use it. If social media companies were to build in more friction or braking mechanisms for mimetic behavior, they would decrease user engagement and ultimately revenue; they have strong financial incentives to *accelerate* mimetic behavior. If two people argue on a social media platform, drawing others into the feud, it's not hard to see who wins: the platform.

Systems of desire, both positive and negative, are everywhere. Prisons, monasteries, families, schools, and friend groups operate as systems of desire. And when a strong mimetic system is in place, it remains in place until it's disrupted by a stronger one.[8]

Few people have experienced the mimetic system of French haute cuisine, but it can shed light on our own. Let's look at how Chef Sébastien Bras got into the system, and how he got out.

Being Watched and Rated

Sébastien Bras's restaurant, Le Suquet, occupies a picturesque hillside on the outskirts of Laguiole, France, about a two-and-a-half-hour drive from each of the three major cities closest to it (Clermont-Ferrand, Toulouse, and Montpellier). Still, the restaurant never has an empty table for lunch or dinner—something that some three-Michelin-star restaurants in the middle of Paris can't claim.

The small town of Laguiole lies in the Aubrac region, a granite plateau that spans more than five hundred square miles of south-central France.

The region is home to some of the richest diversity of plants and wildlife in the country. Equally admired are the town's handmade knives, the sturdy and stately Aubrac cows that roam the hills, and the cheese made from those cows' milk.

I drove from my hotel in the middle of Laguiole, making a gradual ascent to a stretch of quiet road on the outskirts of town. At the bottom of the driveway, a sign reads BRAS. The thin letters are emblazoned on a sign of white frosted glass that is planted in ground covered with wildflowers, grasses, fennel, and many things that you and I might call weeds, all of which are featured on the menu.

Michel Bras, Sébastien's father, built this restaurant on one of the highest plateaus in the area. At the top of the steep driveway, Le Suquet hovers. The restaurant's modern structure, with one asymmetrical corner, is nestled into a hillside. With its floor-to-ceiling windows on all sides, it looks like the observation deck of a well-designed spaceship carrying otherworldly explorers who came in search of foie gras and aligot.

In 1980, long before this newer building on the plateau was built, the elder Bras introduced the world to his *gargouillou*, a dish that contains between fifty and eighty freshly picked vegetables, herbs, and edible flowers from the Aubrac, individually prepared and held together with cured ham cooked in bouillon. Today, focusing on hyperlocal foraged ingredients is trendy; then, it was offering diners weeds for their meal.[9]

Michel Bras was an innovator, but he played inside the lines of the Michelin Guide's star system. His son, Sébastien, would go outside those lines. Very few have ever done so.

When most French chefs open a new restaurant, they wait in trepidation for a Michelin inspector—whom they may or may not recognize as one—to walk through the front door.

On any given day, any one of hundreds of plates that their kitchen sends to the dining room could wind up in front of an inspector. After finishing the meal, the inspector might flash credentials and ask to see the kitchen. Or they might depart anonymously.

The inspector's verdict has life-altering consequences. Having a Michelin star added can make a chef's career and help a restaurant become financially sustainable. Having one taken away can start a death spiral.

Inspectors are kind of like the mysterious characters named "Watchers" in Orson Scott Card's short story "Unaccompanied Sonata." A young boy living in a dystopian, authoritarian society is declared a musical prodigy. He's given strict rules to follow about how he is to develop his talent. When he breaks the rules, a band of anonymous men called Watchers show up, unannounced, wielding sharp knives, to cut off his fingers. Either you play by their rules, or you lose the ability to play.[10]

In 2003, the Michelin watchers swooped in on Bernard Loiseau, a three-star chef in France. They told him they were worried about the lack of inventiveness and artistic direction of his restaurant, implying that he might lose a star. (And Gault Millau, another restaurant guide in France, had recently downgraded Loiseau's restaurant La Côte d'Or from a 19/20 rating to 17/20.) Around this time, after a full day of work in his kitchen, Loiseau committed suicide.

Back in 1994, at the age of thirty-two, British chef Marco Pierre White was the youngest chef ever to be awarded three stars. In 1999, only five years later, he retired. "I gave Michelin inspectors too much respect, and I belittled myself," he explained. "I had three options: I could be a prisoner of my world and continue to work six days a week, I could live a lie and charge high prices and not be behind the stove or I could give my stars back, spend time with my children and re-invent myself."[11] He was the first three-star chef in history to shut down and walk away.

Paris chef Alain Senderens, tired of trying to keep up, closed down his three-star restaurant and revamped it, which kept Michelin temporarily away. "I feel like having fun," he told the *New York Times* in 2005. "I don't want to feed my ego anymore. I am too old for that. I can do beautiful cuisine without all the tra-la-la and chichi, and put the money into what's on the plate."

Each of us has our own version of a Michelin star system. We can easily find ourselves, like a French chef, wanting "stars"—marks of status and prestige, badges of honor. Naming the mimetic forces at work in the systems in which we operate is an important first step toward making more intentional choices.

Tactic 8

MAP OUT THE SYSTEMS OF DESIRE IN YOUR WORLD

Every industry, every school, every family has a particular system of desire that makes certain things more or less desirable. Know which systems of desire you're living in. There's probably more than one.

Entrepreneur and VC Marc Andreessen, in an April 2020 post on his company's website titled "It's Time to Build," wonders how so many Western countries were unprepared—from a production standpoint—for the COVID-19 outbreak in 2020. At one point, there were serious shortages of ventilators, test kits, cotton swabs, even hospital gowns. The complacency and malaise seemed to extend to many other domains, even before the pandemic—to education, manufacturing, transportation. Why were Americans no longer *building* the things of the future? he asked.[12]

The problem is not capital or competence or even a lack of awareness of what's needed. "The problem is desire," Andreessen wrote. "We need to 'want' these things." But he acknowledges that there are forces in place that prevent us from wanting to build the things we need: regulatory capture, industry incumbents, stalemate politics. "The problem is inertia," he continued. "We need to want these things more than we want to prevent these things."

Crippled systems of desire, unable to adapt, have made it so that we gravitate toward the path of least resistance—monetizing YouTube videos of people reacting to other YouTube videos, for instance—and lack the will to build the essential tools needed for human survival and flourishing.

If you understand the systems of desire that color the choices of people around you, you're more likely to see emergent possibilities by daring to look in different directions.

Make visible what is invisible. Mark the boundaries of your current world of wanting, and you'll gain the ability—at least the possibility—to transcend it.

The Tra-La-La and the Chichi

The Michelin Guide is an intermediary, a mediator of desire, from which thousands of chefs seek an imprimatur. Since 1900, when the first Michelin Guide was published, the Michelin brothers have been turning a flywheel of desire.

Michelin employed mimetic marketing. If the company could position itself as a model of desire in regard to which restaurants one should want to go to, Michelin would go from being a company that sells tires to a company that sells desire. They would go from being Compaq to Apple.

At the turn of the century, there were very few cars on the road. Michelin's plan was predicated on the *future desire* of people to drive cars. The creators of the guide, perhaps understanding the reflexive nature of desire, intuited that they could play an important role in generating the very desire they were betting on.

By 1920, the guide had grown into one of the most widely circulated publications in the country. Today, it's one of the most revered things in print.

There is no doubt that the Michelin Guide provides value for millions of people who use it to find an unforgettable dining experience or simply a delicious meal. François Michelin, CEO from 1955 to 1999, was an exemplary leader who embodied a human-centered approach to business. But the guide became something that nobody at Michelin—certainly not in its early days—ever imagined: a limiting and stifling system of desire.[13]

"You get stuck in a dreadful system," Bras says. "If you don't respect the official or unofficial codes and practices of the guide, you risk being downgraded. This is awful for the reputation of the restaurant, for the morale of the chefs, for the entire team. To be downgraded is to fail."

In order to escape, Bras went back to the beginning.

The One Less Traveled By

Truth has a history. We can't know ourselves without knowing the history of our desires.

As we sit in his office conversing, Sébastien Bras, unprompted, recalls his own history. "My room as a kid was located right above the kitchen of my parents' restaurant. I woke up, went to sleep, and spent my afternoons surrounded by sounds from the kitchen: the service, the transport of produce from the market, the stressful moments during working hours, the cooks laughing as they left for the night." His fond memories of the kitchen were formative.

"I also spent a lot of time hiking in the Aubrac with my parents," he says. "When it came time to choose a profession, I thought to myself, 'To be a cook will mean staying here and continuing to have fun.' It was about not wanting to leave my playground."

Bras tested his desire. He went away to secondary school and followed a general education program. "I passed my baccalaureate exams in economics to make sure it was this job—becoming a chef—that I wanted. I wanted to make sure I wasn't doing it to take the easy way out."

He received the confirmation that he needed—the desire to be a chef endured and grew throughout secondary school. It would continue to be tested time and time again as he embraced his role, but he could rest confident knowing that he had kicked all of the tires first.

Bras committed himself to his family's restaurant. He loved cooking, and he knew what he was getting himself into. "My objective back then was to help my father get the three Michelin stars and keep them," he says.

When Sébastien joined his father, Le Suquet had only two Michelin stars. Getting the third one was critical. Michelin awards three stars only to the best restaurants in the world—places the guide says are worth "making a special journey" for.

Because Le Suquet is located in a remote place, the restaurant relied more heavily than most on this distinction. The guide's stars would allow the Bras family to establish a base of business and make the restaurant known outside the region.

Le Suquet achieved the third star in 1999. For the next ten years the restaurant grew in prestige and accolades.

But shortly after Sébastien officially took over the restaurant from his

Tactic 9

PUT DESIRES TO THE TEST

Don't take desires at face value. Find out where they lead.

Sit with competing desires and project them into the future. Let's say you have two competing job offers: Company A and Company B. If you have two days to make the final decision, spend one day with each company in your imagination. On the first day, imagine with as much detail as possible that you're working at Company A and fulfilling the desires that come along with that position—maybe it's living in a new city, interacting with smart people, and being closer to your family. Pay close attention to your emotions and what's going on inside your gut. The next day, spend the entire day doing the same thing, except at Company B. Compare.

The ultimate way to test desires—especially major life choices such as whether to marry someone or whether to quit your job and start a company—is to practice this same exercise but to do it while imagining yourself on your deathbed. Which choice leaves you more consoled? Which choice causes you more agitation? Steve Jobs, in his 2005 commencement speech at Stanford, noted, "Death is very likely the single best invention of life. It is life's change agent. It clears out the old to make way for the new." The deathbed is where unfulfilling desires are exposed. Transport yourself there now rather than waiting until later, when it might be too late.

father in 2009, he and his wife began to experience the Michelin system not as an exhortation to excellence but as a source of pressure and conformity.

Many elite chefs feel their creative ambitions constrained after they gain entry into the pantheon of three-star restaurateurs. Everything becomes subordinated to keeping the stars. They become more risk-averse. The Michelin inspectors have expectations. Why risk something that the inspectors might not like?

Chefs know that certain items need to be on the menu: locally sourced foods, elaborate cheese carts, multiple dessert options, an extensive wine list. World-class sommeliers and an army of highly trained servers and staff (which come at a cost) are also expected.

And the meal is only the beginning. If the restaurant isn't in a major city, it is hard to be considered for the top Michelin star rating without having guest rooms on the property. Chefs have to also be hotel operators to compete. Maison Bras, part of the Relais & Châteaux network, features eleven guest rooms and two apartments in a complex attached to the restaurant.

As we converse in his office, we finally reach the point where I ask Bras why he made the decision to give back his stars. He tells me that Michelin tried to set itself up as "both judge and jury."

"Six or seven years ago," Bras says, "they sat in my office to explain their new marketing strategy to me. They wanted me to buy various commercial services and tools." Every restaurant was ostensibly free to use or not use Michelin's new tools, but Bras didn't like it. "Michelin had the ability to judge and destroy anyone's reputation—and at the same time sell them marketing tools. To me that was not acceptable."

He was playing a game with no end, and he was exhausted. "Eventually you have to stop. You end up working no longer for yourself or your clients, but according to the so-called expectations of the guide."

He began to ask: "Did I choose this career so that the reputation of my company would depend on another institution? Do we want to live another fifteen years of stress or pressure?"

Modeling a New Mindset

On Father's Day 2017, while mountain biking in the Aubrac, Bras realized that he wanted something else more than he wanted status in the Michelin Guide. He wanted to create new dishes that shared the terroir of his region with others, and it didn't matter what the Michelin inspectors thought.

Bras told me that he hadn't felt entirely free to try new culinary experiments for a long time. To be able to creatively express his love for the Aubrac through its food had been his primary and enduring desire for as long as he could remember. He had forgotten how much it meant to him. Now he wanted to reignite it.

By the time he returned from his bike ride, he'd made up his mind. Though he hadn't talked to any other chefs about withdrawing from the Michelin Guide, nor discussed the matter with anyone but his wife, he knew it was time to move on. Bras called the international director of the guide, Gwendal Poullennec, and requested that Le Suquet be removed.

In the 120-year history of the guide, Bras's decision was unprecedented. Chefs had tried to opt out of the system by closing, relocating, or overhauling their concepts. But Bras wasn't changing anything about his restaurant. Same menu, same prices. He simply didn't want Michelin to come back.

The response was courteous but confused. Bras didn't receive any indication as to whether his request would be heeded.

In September 2017, Bras posted a video on Facebook publicly requesting removal. "Today, at forty-six years old, I want to give a new meaning to my life . . . and redefine what is essential," he says in the video, dressed in his chef's whites with the rolling landscape of Laguiole behind him.[14]

"I didn't give them a choice," Bras tells me. He wanted to use the power of social media to mobilize public sentiment in his favor. It worked. Within a week, the video had over a million views.

Bras heard nothing from Michelin for months. It's not hard to imagine the questions on their mind, though. What if more chefs opt out? What would that mean to the long-term value of the brand? What if Bras provides a precedent? A model?

In February 2018, when the new guide was released, Bras discovered that Le Suquet was not in it. He was free.

"What was that year like?" I ask.

"*Parfait*."

That year Bras spent more time with his wife and two children. He carried less weight on his shoulders. He felt free to create and to play. He had drawn a line at what "more" means to him.

Was Bras's decision easier because he had already achieved three stars? Probably. He certainly couldn't be accused of sour grapes.

The term "sour grapes" was popularized in one of Aesop's fables. A fox sees a beautiful cluster of ripe grapes hanging from a high branch. The grapes look ready to burst with juice. His mouth begins watering. He tries to jump up and grab them, but he falls short. He tries again and again, but the grapes are always just out of reach. Finally, he sits down and concludes that the grapes must be sour and aren't worth the effort anyway. He walks away scornfully. By calling the grapes "sour," the fox invented a narrative in his mind to ease the pain of loss.

If you accept this notion uncritically, then you might believe that one can't legitimately despise rich people without first being rich, or scorn Ivy League schools without having gained admission to one, or reject the desire for three Michelin stars without first having earned them. To do so would be self-deception, resentment, weakness.

Don't believe that a person has to buy into and play a mimetic game and *win* before they can opt out of it with a clear conscience.

If you decline an invitation to be on the reality TV show *The Bachelor*, rejecting it as a silly charade, does that mean that it's sour grapes? Could you only criticize the show after you've won? Of course not. "Don't knock it till you try it" is a sophomoric argument.

Girard recognized that resentment is real—and that it happens primarily in the world of internal mediation (Freshmanistan), when we are inside a system of desire without social or critical distance from it.[15] But only the worst kind of cynic believes that every renunciation necessarily has something to do with resentment.

Would it have been harder for Bras to renounce the stars if he had only earned two and was the verge of earning a third for the first time? Almost certainly. It's the challenge that each of us who has ever been a part of a mimetic system—that is, everyone—faces.

In adulthood, we are free to pick some of the systems of desire that we are a part of and alter the nature of our relationship to others. The earlier we exercise our agency in the process, the easier it is.

One detail of Aesop's fable is rarely mentioned: the fox was alone. He had no mimetic forces operating on him. Had there been just one more fox salivating over the grapes, he wouldn't have been able to call them "sour" so easily. Had there been a group of foxes that wanted them, it would've been nearly impossible. But on his own, the fox was a Romantic Liar.

If you've made it this far in the book, you no longer have that luxury.

We can pretend that a good thing is bad, and maybe even that a bad thing is good—but it's far more difficult for us than it was for the fox because we have to contend with other people who are signaling value, both good and bad.

Sébastien Bras tasted the grapes, and the grapes were sour. Do you need to taste to believe?

Bras was able to extract himself because he changed his relationship to the game.

"We live in a society where we are always being asked for more," Bras says to me. "To be stronger, to go higher, to get bigger numbers, ever greater and ever higher. But I think there's a deep desire in people to reconnect with true life values. Values that we tend to forget sometimes." For Bras, those values centered around his family and his desire to create and share the food of the Aubrac region without fear of reprisal.

If Sébastien could be a model of desire for gaining three stars, then maybe he could be a model for renouncing three stars, too. "I think my decision revealed the deep desires of those chefs who thought, 'Wow, someone who has dared to say no to the system? Maybe I can, too. Now maybe I can live my life, too.'"

For nearly a week after he announced his decision on Facebook, his phone was ringing off the hook from seven in the morning until ten in the evening. Bras noticed that people tended to respond in one of two ways to his decision. "I talked to a number of three-star chefs who perfectly understood why I did what I did," he says. "But there are chefs with one or two stars whose sole objective is to obtain an additional star. They didn't understand my decision."

In February 2019, exactly a year after learning of his initial victory, Sébastien received a phone call. "It was a Sunday evening, around eight o'clock, the day before the 2019 guide was going to be released," he told me. On the other side of the line was Poullennec. "He informed me that I was being reintegrated into the 2019 guide—and with two stars."[16]

"And what was your reaction?" I asked him.

"I laughed," he said. "I laughed a lot."

Chapter 6

DISRUPTIVE EMPATHY
Breaking Through Thin Desires

VIVIAN WARD: Look, you made me a really nice offer. And a
few months ago, no problem. But now everything is different, and
you've changed that. And you can't change back. I want more.
EDWARD LEWIS: I know about wanting more.

—*Pretty Woman*

The only true voyage of discovery . . . would be not to visit strange
lands but to possess other eyes, to behold the universe through the
eyes of another, of a hundred others, to behold the hundred uni-
verses that each of them beholds, that each of them is.

—Marcel Proust

Dave Romero (a pseudonym) had a business card that said "Customer Relations Specialist." He showed up at my house on a Friday morning after
failing to find me in my office near downtown Las Vegas. Dave was from
what they call "old" Vegas, before the casinos with fake volcanos and Mich-
elob Ultra, when it was a seedy Wild West.

I worked with over a thousand suppliers at my company. I had personal
relationships with a hundred of them. Back when I first started the company, I had long, rollicking conversations with all of the sales reps. But as we
grew, my time got diverted to other things. Now most of my suppliers were

a black box to me—they sent products, we sold the products on our website, and thirty days later we paid the invoice.

Until one day I couldn't pay.

It was late 2008, after my deal with Zappos fell through. I'd already maxed out my credit to keep the company afloat while I figured out next steps. I had experienced relief with the implosion of the Zappos deal because it unburdened me from borrowed desires—but it was short-lived, because I was now facing the unpleasant reality of managing a flailing company.

In an effort to buy myself some time, I made a list of the suppliers I would pay first. Companies that were strict about getting paid on time—the most likely ones to come after me if they were not—were at the top of the list; companies that I knew had more relaxed people in charge of accounts receivable were near the bottom. There was one major problem, though. The list was made up entirely of companies that I had personal relationships with. The black-box companies didn't make the list because I didn't know anything about them, so I had no basis on which to evaluate them.

One company, Fyre Pharmaceuticals (not its real name)—for which Dave Romero specialized in customer relations—had been in the black box. My decision to exclude it from my list of priority payments would have been different had I known that the company's founders were rumored to have connections with organized crime, that they were said to be involved in gun trafficking, and that one of my competitors mysteriously vanished after crossing them.

I would learn these things by asking around in the industry after Dave Romero first showed up in my life. By then it was too late.

Dave Romero had a thin ponytail and a sallow face. His narrow eyes and deep crow's feet gave the impression that he could read your soul. He walked proudly, with confidence. I imagined that he'd single-handedly and meticulously broken the fingers of five Viet Cong sympathizers with a beer bottle in a bar in Saigon and still, because of that, wakes up every morning, looks in the mirror, and thinks, *I'm Dave fucking Romero.*

He showed up at the front door of my house at seven o'clock in the morning as I was getting ready to take my dog for a walk. When I heard the doorbell, I thought it was the Mormon missionaries again. But they don't come at seven a.m. What I found instead was Dave Romero.

I had interacted with Dave three times before. The first was an unpleasant phone call during which he informed me that my payment was late; offended, I snapped back with a full-throated defense of my excellent payment history. The second was an unannounced visit to my office, where he confronted me and told me that he wasn't a patient man; he threw a pair of dice on my desk and left. The third was when he showed up at a local bar during Sunday football—no idea how he'd known I was there—and told me that he "means what he says" while beating one of his fists against his palm like he was tenderizing meat; the bouncers escorted Dave out. As he left, he made the figure of a pistol with his thumb and forefinger, and pointed it at me.

This was now the fourth visit from Dave, and I didn't know what it meant.

He seemed different this time. He began with small talk. He asked me how I was doing. He commented on the weather. Is this how it goes? Was it the happy-go-lucky backslapping before a guy gets whacked on *The Sopranos*? His friendliness made me suspicious.

I bumbled nervous replies. I stood taking up as much space in the front door as I could to prevent my dog Axel, who was standing behind me, hair on end, from slipping out. Dave was standing too close, seeming like he wanted to slip in.

He drew even closer and lowered his voice. "Can you please, um, quick handle this bill by Monday morning so I don't have to . . . come back?" He spoke quietly, calmly, and courteously while he twisted one of the gaudy rings on his left hand.

There was no way that I could pay him that soon.

Before I had a chance to respond, he continued: "And, oh, hey, I hear you're having a big company barbeque here tomorrow night."

It was true. I hosted a monthly party at my house with rolling invitations to people at my company. This time, though, I had invited everyone. I worried that it might be our last rendezvous if things didn't turn around.

But how did Dave Romero know about it?

"I . . . and so but . . . how did you . . ."

"Mind if I come?" he asked.

It didn't seem like a question. I was growing increasingly confused and nervous. I just wanted Dave off my porch. "No, I mean yeah, sure, people start showing up at seven, you can stop by." The words came out of my mouth. I'd never refused a request to come to one of my parties—certainly not to anyone's face. I didn't know how.

And now I had invited a hit man into my home.

Dave showed up with a bottle of Four Roses Single Barrel Bourbon and insisted I enjoy it with no more than one ice cube. The party was a hit. When the last dregs had been drunk and the last ember on the grill had died, nobody wanted to go home. Including Dave Romero.

Dave sat out back with a few of us around the fire pit. I'd already sweated through two shirts and thought I might need to do another outfit change by the end of the night.

He had been coy about his true reason for being there. I had not told anyone else about my interaction with him from the day before. Only a few people in the company besides me had interacted with Fyre Pharma in their job roles. Most people had no idea who he was. "I work with Luke," he said when asked. And that was that. Nobody was talking shop, and nobody cared. From time to time I invited an outsider to these barbeques, so people just assumed Dave was another one of my weird friends.

Dave, who had been mostly silent as a few of us sat around the fire, spoke up during a break in the storytelling. "What's the unhealthiest thing you've ever done?" he asked.

I admitted that I'd drunkenly eaten a double cheeseburger sandwiched between two Krispy Kreme glazed donuts in college. Paul said he had a fair amount of unprotected sex while living in Thailand. Jessica said she used to do whippets. Her husband, Tom, shared that he had secretly reverse-mortgaged their home to place risky bets in the stock market while their first baby was on the way.

"I killed a man," Dave said.

I stared into the fire watching the flames lick the logs, wondering if I'd heard right. I felt Dave's eyes on me and everyone else's eyes on him. The fire was dying down. I tossed in a leftover marshmallow and watched it burn.

Dave scooted up in his seat and leaned forward on the balls of his feet. "What would you all dream of doing if you stopped worrying so damn much about money?"

"I don't know," I said. "I'd have to stop worrying so much about money first."

Over the next hour, Dave asked increasingly personal questions of everyone around the fire—questions that are usually answered in a eulogy, not at a casual work gathering. What's the most fulfilling thing you've ever done? Whom have you deeply loved? Where do you go when you want to numb the pain?

Dave was vulnerable, so others were, too. Dave told us that he wanted to use the last decade of his life—which he was convinced he was in—to fulfill some of the promises he'd made to people, like taking his nephew skydiving, visiting prison inmates once a month, and getting out of his current line of work.

It was after midnight when people started to filter out. Dave was one of the last to leave. I shook his hand and told him that I'd be in touch so that we could handle everything. He laughed and put his hands on my shoulders. "You're okay, Luke, you know?" He slapped my back and stumbled out the front door to catch a cab.

I found out later that week that Dave suffered a heart attack and died. Someone at Fyre Pharma told me that Dave was not a hired hand but a partner, and that he had said that we were "reconciled." I never heard another word from them.

What happened that night is something I now recognize as *disruptive empathy*.[1] The cycle of conflict that stems from unchecked mimesis—like that of a debt collector and a debtor, each responding mimetically to the aggression of the other—was derailed. There was an unexpected breaking in of empathy, something that transcended the moment.

Fear, anxiety, and anger are easily amplified by mimesis. A colleague sends me an email that seems curt or disrespectful, I respond in kind; my friend raises his voice in an argument, I raise mine back; and passive aggression spreads like wildfire, beyond two people and through an entire organizational culture.

René Girard uses the example of a handshake gone wrong to illustrate how deep-rooted mimesis is—and how it explains things we usually ascribe to simply being "reactionary." There's nothing trivial about a handshake. Say that you extend your hand to me, and I leave you hanging. I don't imitate your ritual gesture. What happens? You become inhibited and withdraw— probably equally as much, and probably more, than you sensed I did to you. "We suppose that there is nothing more normal, more natural than this reaction, and yet a moment's reflection will reveal its paradoxical character," writes Girard. "If I decline to shake your hand, if, in short, I refuse to imitate you, then you are now the one who imitates me, by reproducing my refusal, by copying me instead. Imitation, which usually expresses agreement in this case, now serves to confirm and strengthen disagreement. Once again, in other words, imitation triumphs. Here we see how rigorously, how implacably mutual imitation structures even the simplest human relations."[2]

This is how negative mimetic cycles start. We are not condemned to them, though.

In this chapter we'll learn a specific approach for getting to know people at their core, which reduces the possibility of cheap mimetic interactions. The approach involves sharing and listening to a particular kind of experience: stories of deeply fulfilling action. Knowing and relating to these stories produces empathy and a greater understanding of human behavior.

A negative mimetic cycle is disrupted when two people, through empathy, stop seeing each other as rivals. Dave changed my way of thinking and my reactionary impulses by modeling something different—a core desire that is common to every person, but which often goes unfulfilled: to know and be known by others.

We can imitate a management strategy, or we can imitate empathy. The first is a framework; the second is a process. The process that I'll outline here has to do with becoming more attentive to the humanity—of others, and our own—that goes beyond any framework.

The Problem with Sympathy

"I'm sympathetic to the cause." You've probably heard these words before. They're common—and that's partly because sympathy is far easier to practice than empathy.

The word "sympathy" shares the same root as "empathy"—they both come from the Greek word *pathos*, which roughly means "feeling" or something that appeals to the emotions (according to Aristotle's use of the word). The difference between the two words is in the prefix. Sympathy starts with *sym-*, meaning "together." Sympathy means "feeling together." Our emotions fuse with those of the person we sympathize with. We see things from their perspective. A certain degree of agreement is implied.

Sympathy can be easily hijacked by mimesis. Have you ever been part of a group that begins a conversation about something and rapidly coalesces around some form of agreement—maybe about politics, a business decision, or what looks good on a menu? You find yourself nodding along, smiling, maybe even agreeing out loud. But a few minutes later, or when you get home later that night, you think, *Hold on . . . do I* really *agree with that?*

Empathy feels different. The *em-* in empathy means "to go into." It's the ability to *go into* the experiences or feelings of another person—but without losing self-possession, or the ability to maintain control over our responses and to act freely, out of our own core.

Real empathy is undertaking an intentional journey—like the divers who entered the Tham Luang cave in Thailand in 2018 to rescue a trapped soccer team. They entered that cave of their own volition. They were in control of themselves as they made their way to the trapped children, hyperaware of their surroundings and their responses in order not to get lost or perish.

Empathy is the ability to share in another person's experience—but *without imitating them* (their speech, their beliefs, their actions, their feelings) and *without identifying with them* to the point that one's own individuality and self-possession are lost. In this sense, empathy is antimimetic.

Empathy could mean smiling and giving a cold bottle of water to people collecting signatures for a petition you would never sign—because it's a sweltering day and you know what it's like to be that hot, and you also know what it feels like to be that passionate about something you care about. It would not entail empty platitudes or white-lie niceties that we often say to people with whom we disagree; rather, it means finding a shared point of humanity through which to connect without sacrificing our integrity in the process.

Empathy disrupts negative cycles of mimesis. A person who is able to empathize can enter into the experience of another person and share her thoughts and feelings *without necessarily sharing her desires.* An empathetic person has the ability to understand why someone might want something that they don't want for themselves. In short, empathy allows us to connect deeply with other people without *becoming like* other people.

Recall that in a mimetic crisis, everyone starts to become like everyone else. They lose self-possession and freedom. The prolific letter-writer and

Trappist monk Thomas Merton noticed this was happening to him during his college years at Columbia University. Later in life, he wrote: "The true inner self must be drawn up like a jewel from the bottom of the sea, rescued from confusion, from indistinction, from immersion in the common, the nondescript, the trivial, the sordid, the evanescent."[3]

Empathy allows us to interact with others without sacrificing those jewels of our inner selves, without getting swallowed up by the waters. It helps us find and cultivate *thick* desires—desires that are not hyper-mimetic, desires that can form the foundation for a good life.

Thick Desires

Discovering and developing thick desires protects against cheap mimetic desires—and ultimately leads to a more fulfilling life.

Thick desires are like diamonds that have been formed deep beneath the surface, nearer to the core of the Earth. Thick desires are protected from the volatility of changing circumstances in our lives. Thin desires, on the other hand, are highly mimetic, contagious, and often shallow.

I wish I could say that desires necessarily become thicker as we age, but that's not always the case. At least it doesn't happen without intentional effort. We've all met older people who realized too late that their desires were thin—for example, a person who looked forward to retirement for decades only to find out that attaining it left them unsatisfied. That's because the desire to retire (not widely adopted until after World War II, by the way) is a thin desire, filled with mimetically derived ideas about the things one might do, or not do, in this ideal state. The desire to invest more time with family, on the other hand, is a thick desire—and the proof is that a person can start to fulfill it today and continue to fulfill it into retirement. It grows with compound interest over many years. It has time to solidify.

The distinction between thick and thin desires can't easily be made based on feelings alone. Desires feel very strong when we're young—to make a lot of money, date a person with certain physical attributes, or become famous. The feelings are often more intense the thinner a desire is. As we get older, many of our adolescent feelings of intense desire fade away. It's not because we realize that some of the things we wanted are no longer attainable. It's because we have more pattern recognition ability and so can recognize the kinds of desires that leave us unfulfilled. As a result, most people *do* learn to cultivate thicker desires as they age.

But the tension between thick and thin always remains. Every artist has experienced it. They may have had a lifelong desire to tell the truth, to make art that expresses something important. Yet they have a competing desire to sell their work in the marketplace, to be accepted, to be praised, to get reviews, to stay on top of trends that can change from year to year, month to month, day to day. The latter are superficial desires that, if allowed to accumulate, can completely obscure the thick ones.

Sometimes it takes a particular event to shake those thin desires out of us.

Shaking the Dust

When the sale of my company to Zappos fell through in 2008, I was forced to ask myself: why had I wanted to start this company in the first place? I identified at least three thick desires that had been obscured, covered up, and abandoned for thin ones.

First: when I started out, I could not have cared less about whether anyone even knew my name so long as I created a company that created value in the world. How, then, did I become concerned with *prestige*? I craved the recognition of having won some award or prize or achieved a certain number of followers—desires that had been largely foreign to me a few years prior. But since my peers craved recognition, I began to want it, too. I started vying to make a "Best Places to Work" list and other spurious accolades.

The word "prestige" comes from the Latin *praestigium*, which means "illusion" or "conjurer's trick." (The 2006 movie *The Prestige*, which depicts the mimetic rivalry between two magicians, was well titled.) People seek professional prestige—respect or admiration for their talents—without realizing that the pursuit of prestige is the pursuit of a fata morgana.

Within a few years of starting my first company, I spent more time looking sideways than I did forward. I was looking for ways to measure success and finding them everywhere. The kid at the coffee shop with a higher-end laptop. The founder with more prestigious VC backing. The entrepreneurs who didn't seem like they were trying so hard, for whom success seemed to come naturally. I secretly resented all of them.

The old-fashioned word for that experience is "envy." "I think the reason we talk so much about sex is that we don't dare talk about envy," Girard said.[4]

Envy is an engine of destructive mimetic desire, and there are few things to stop it because it operates underground. Prestige is measured relative to what we perceive someone else has that we lack, so it's a breeding ground for envy.

Entrepreneurship has many recognized occupational hazards, from mental health risks to burnout to substance abuse and financial instability. None is so conspicuously absent from public discourse as envy.

Second, I went from wanting to carve out my own lifestyle—one of the greatest perks of entrepreneurship—to conforming my lifestyle to the pattern that other entrepreneurs modeled.

In the beginning, when I quit my job in finance to start a company, I wanted a lifestyle with clear boundaries and balance. I wanted to read for an hour every night, to take long walks with my dog, to spend more time with my friends, to be in a loving relationship. But as CEO of my start-up, I found myself working eighty-hour weeks and disregarding all boundaries and balance. What happened?

The lifestyle in Silicon Valley, and in the start-up world in general, is largely a function of mimesis. Not everyone moved to Menlo Park and decided to wear logoed hoodies and Vans at the same time. Not everyone decided to start sending trite, unimaginative, prosaic emails in all lowercase letters, feigning busyness and self-importance, at the same time. (Here's one way to be anti-mimetic: when you receive one of those emails, respond with something respectful, thoughtful, beautiful.)

In my case, I was infected with the Zappos culture mania. Tony Hsieh spoke about the Zappos culture as though it were a cargo cult—if you simply follow the same recipe, you'll build a successful culture.[5] Financial success comes along with it. It wasn't long before my company's office started looking more like the Zappos office—weird stuff hanging on the walls, quirky celebrations, a library in the visitor's lounge containing the canon of business books. I was at happy hours practically every night. I felt like I wasn't keeping up and like I wouldn't be considered a "culture fit" if I opted out.

Third, I went from craving classical wisdom to consuming memes and tweets and tech news—which led to my imitating ideas without knowing it. I knew more about what blogger Gary Vaynerchuk had to say about happiness than Aristotle. The ecosystem that I lived and worked in seemed to be growing more homogenous by the day. I might have had the courage to stand outside this system of ideas, but how could I? I didn't know anything outside the system.

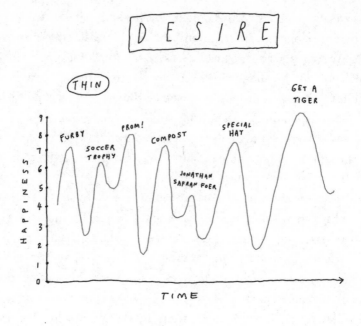

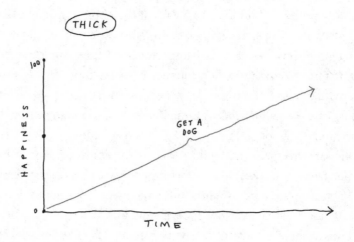

When I reflected seriously on the ideas that dominated my world, I found them shallow. What had happened to the desire, kindled in me early in life, to probe the ideas that had stood the test of time?

I had to change something. My conversation with Dave Romero around the firepit, especially the regret that I heard in his voice, convinced me that most of the desires I entertained were thin and fragile. They were capable of being blown away at any time like dust in the wind. They weren't a solid foundation on which to build a life.

Shortly after receiving news of Dave's death, I started to wind down my company's operations. Not necessarily because of Dave, but because it seemed like something I had to do in order to complete the gradual transformation I was undergoing during this period.

I realized that one of my deepest desires was to explore life's Big Questions—to understand humanity at a deeper level, starting with myself.[6]

I wanted to do that more than I wanted to work ninety-hour weeks to streamline my warehouse operations, find ways to conserve cash, and eventually save a company that I no longer wanted to lead.

I decided to take a three-month hiatus from the start-up world so that I could reorient myself—primarily, reorient my desires—before deciding what to do next. Those were the first three months of the rest of my far less mimetic life.

Fulfillment Stories

Let's say I've convinced you that it's time to put aside thin desires and focus on the more anti-mimetic, rooted, solid ones. The hard work will have just begun. Thin desires aren't so easily dismissed, and thick desires are not something that you can self-generate out of thin air. They take time—months and years—to develop.

The best place to start is with the thick desires you probably already have. They're not always easy to identify. Thick desires are hidden beneath the fleeting and impulsive desires that dominate most of our days. The American author and educator Parker Palmer writes, "Before I can tell my life what I want to do with it, I must listen to my life telling me who I am."[7]

The approach that I'll outline here is anthropological, philosophical, practical, even spiritual. I like the definition of spirituality by Rabbi Jonathan Sacks, who writes that spirituality is simply "what happens when we open ourselves to something greater than ourselves." He continues: "Some find it in the beauty of nature, or art, or music. Others find it in prayer, or performing a mitzvah, or learning a sacred text. Yet others find it in helping other people or in friendship or love."[8] It could be described as a sense of connectedness to self and to others and to the universe.

As we've seen so far, desire is social. Desire *is* connectedness. So my hope is that even if you don't think of yourself as spiritual, these approaches will be helpful because they are grounded in a fundamental truth about what it means to be human: we are not entirely our own, but exist in a web of relationships connected by desire.

One approach I recommend for uncovering thick desires—the one I'll focus on here—involves taking the time to listen to the most deeply fulfilling experiences of your colleagues' (or partners', or friends', or classmates') lives, and sharing your own with them. The more we understand one another's stories of meaningful achievement, the more effectively we understand how to work with each other: what moves and motivates others, what gives them satisfaction in their work.

It seems simple, but nobody does it. Ask yourself: How many people do you work with who could name even one of your most meaningful achievements and explain why it was so meaningful to you?

A key goal of this exercise is identifying core motivational drives. A motivational drive is a specific and enduring behavioral energy that has oriented you throughout your life to achieve a distinct pattern of results. You may be fundamentally motivated, for instance, to *bring control*, to *evoke recognition*, or to *overcome obstacles*. Because most of us have never thought seriously about the nature of our motivation, we lack the language to be able to describe our core motivational drives with much precision. This exercise gives us that ability.

Core motivational drives are enduring, irresistible, and insatiable. They are probably explanatory of much of your behavior since the time you were a child. Think of them as your motivational energy—the reason you consistently gravitate toward certain types of projects (team versus individual, goal-oriented versus ideation) and activities (sports, arts, theater, forms of fitness) and not others.

There are patterns in your motivation. If you can put your finger on what specifically they are, then you will have taken a major step toward understanding your thick desires. The best way to uncover the patterns is by sharing stories.

The storytelling process that I use involves sharing stories about times in your life when you took an action that ended up being deeply fulfilling. Today it's one of the first questions that I ask in any job interview because it helps cut through the thin stuff and goes straight to the essence of the person. "Tell me about a time in your life when you did something well and it brought you a sense of fulfillment," I ask.

I have seen this simple question transform interactions between individuals and entire communities. When stories are shared between two people who know how to listen well, the experience transports both the storyteller and the listener to a time when desire led to extraordinarily fulfillment. That's why sharing those stories is a joyful experience.

A Fulfillment Story, as I call it, has three essential elements:

1. **It's an action.** You took some concrete action and you were the main protagonist, as opposed to passively taking in an experience. As life-changing as a Springsteen concert at the Stone Pony might have been for you, it's not a Fulfillment Story. It might be for Bruce, but not for you. Dedicating yourself to learning everything about an artist and their work, on the other hand, could be.

2. **You believe you did well.** You did it with excellence, you did it well—by your own estimation, and nobody else's. You are looking for an achievement that matters *to you*. If you grilled what you think is a perfect rib-eye steak the other night, then you did something well and achieved something. Don't worry about how big or small the achievement might seem to anyone else.

3. **It brought you a sense of fulfillment.** Your action brought you a deep sense of fulfillment, maybe even joy. Not the fleeting, temporary kind, like an endorphin rush. Fulfillment: you woke up the next morning and felt a sense of satisfaction about it. You still do. Just thinking about it brings some of it back.

Such moments of profound meaning and satisfaction matter. They reveal something critical about who you are.

"Actions follow being," wrote Aristotle twenty-three centuries ago. He meant that a thing can only act according to what it is. We can know something about the essence of a thing based on its actions. But in the case of humans, we also need insight into the interior dimension of action: What was the person's motivation for taking it? What were the circumstances? How did the action affect them on an emotional level?

Imagine three artists standing shoulder to shoulder on a plateau in Zion National Park painting the same sunset. One wants to hone his painting skills for a competition; another wants to give the finished painting to her husband on their anniversary because they had their first date in the park; the last wants to preserve the sheer beauty of the landscape in her memory. From the outside, the artists appear to be doing exactly the same thing. From the inside, each artist is doing something very different.

We can mostly understand the actions of cats and dogs from the outside looking in. But people are different: knowing something about the interior life of a person is necessary to understand why they do what they do and what it means to them. Fulfillment Stories get at the heart of action by looking at it from the inside out. Fulfillment Stories ask, "But *why* did that action mean so much *to you*?"

The question and answer kick off a positive mimetic cycle. You tell one of your Fulfillment Stories. I listen with empathy to what you are telling me and reflect back to you what I heard and saw and felt in your story. Then you do the same for me. Empathy imitates empathy, heart speaks to heart.

It was about ten years ago that I was asked to share one of these stories for the first time, when a friend of mine who specializes in narrative psychology took me through the process. Every time I told a Fulfillment Story, another rose to the surface. As I dove deeper into my past, I uncovered stories I hadn't thought about in a long time. Not only that—I hadn't even recognized them as stories of fulfilling action at the time.

The no-hitter that I pitched in Little League.

Getting my first company off the ground.

Writing every day for thirty days.

Some surprised me:

Making homemade pierogies for dinner using my grandma's recipe.

Inventing an orange-peeling machine in my fifth-grade science class.

Debugging my start-up's website by teaching myself PHP and MySQL.

A few of these things were probably not recognized as "achievements" by anyone else around me at the time. They were achievements *to me*—they gave me a great deal of satisfaction. As I described them, a pattern of thick desires began to emerge.

Motivational Patterns

The approach that I'm describing can be practiced by anyone, anywhere, with nothing more than goodwill and empathy.

With that said, one organization that I've worked with for many years has codified common motivational patterns into an assessment (trademarked under the name the Motivation Code, or MCODE) that identifies and defines twenty-seven separate themes.[9] The assessment asks people a series of questions about each of their Fulfillment Stories and then finds patterns based on what people rated as the most deeply satisfying aspects of their achievements.

Each person possesses a mix of core motivational drives. The key is learning how they work together and how some of them come into play more than others in certain situations.

Following are three of the twenty-seven motivational themes, as defined in the MCODE, with examples that show how they play out. If you're curious to know the rest, I've provided resources in Appendix C. The name of each theme is in bolded **ALL CAPS** defined in the MCODE.

> EXPLORE: People motivated to EXPLORE want to press beyond their existing limits of knowledge and experience to discover what is unknown or mysterious to them.

My friend Ben (that name and the names of my other friends here are pseudonyms) loves couch-surfing in new countries and learning about the language and food of the people he's visiting. He was once so fascinated by the many varieties of spices at a market in Turkey that he spent hours tasting and learning about them, WhatsApping me a running commentary. Once the spices were no longer mysterious, he moved on to exploring other things: craft cocktails, seventeenth-century French literature, cryptocurrency.

The range of interests and the speed at which Ben can move from exploring one thing to another may seem dilettantish. But some people are strongly motivated to **EXPLORE** like this—and that's okay. In fact, every core motivational drive is inherently good. It's simply how we're wired.

There's a shadow side to every core motivational drive, though. Because Ben now knows that he's fundamentally motivated to **EXPLORE**, he is aware when he is tempted to become distracted by new possibilities and not follow through on existing commitments. He is now intentional about channeling his motivational energy to productive, value-creating things. He's now writing a book about his travels. (The popular travel writer Rick Steves is, if I had to guess, also fundamentally motivated to **EXPLORE**.)

It's important to note that Ben is *not* motivated to **MASTER** any one particular thing. When Ben and I traveled in Italy together, we both learned enough Italian to get by. Our mutual friend Alex was bent on mastering the language. While we dabbled in it and amused ourselves with new words, Alex tucked himself away in his room with an Italian-language copy of the Pinocchio story and would not be satisfied until he knew how to use every word in a sentence. That's because Alex *is* fundamentally motivated to **MASTER**.

> **MASTER:** A person motivated to MASTER wants to gain complete command of a skill, subject, procedure, technique, or process.

Unlike my friend Ben and me, Alex wasn't content with the over-the-top praise native Italians lavished on us for trying: *Parli molto bene l'italiano!* (You speak Italian very well!) He wasn't happy until he was able to read classical Italian literature and describe what he needed at the farmers market in Rome's Campo dei Fiori.

Alex went on to pursue a doctorate in physics. When I introduced him to the music of Snail Mail in Ellicott City, he devoured her music until he knew every lyric and could re-create her songs on his guitar. He has few interests, but he goes deep into the ones that he has. Indie rock is one of them.

Alex was never motivated to display his mastery to others in any way. He has no social media accounts. When he gained proficiency with the guitar, he had no desire to start a band. For him, mastery is its own reward.

Another friend, Lauren, has a core motivational drive distinct from either Ben's or Alex's. She loves to write nonfiction because her fundamental motivation is to **COMPREHEND AND EXPRESS**.

COMPREHEND AND EXPRESS: A person with this core drive wants to understand, define, and then communicate their insights in some way.

Lauren feels stymied and loses motivation without an outlet to express new learning. If she reads a book, she is compelled to review it on her blog. If she can't find a way to express it, her new knowledge feels to her like it has been lost or at least not fully developed. Her understanding is clarified in the act of expression.

This is true for her not only in the realm of ideas, but also with experiences. When she tries a new cuisine, she is not content to eat out. She will try her hand at making sushi or paella. This is more than merely a learning style, it's a core motivational drive—core because it applies to absolutely every dimension of her life. It comes out in her marriage (she tries to comprehend family dynamics by thoughtfully listening to everyone, and writes letters to members of her family expressing what she understands their gifts to be); how she approaches a crisis at work (she is an expert moderator of debates, drawing out and communicating key insights); even how she approaches fitness (it wasn't enough for her to just do yoga; she had to become a yoga instructor).

Gaining an insight into your core motivational drives will allow you to understand why some activities are always deeply engaging for you while others inevitably disengage you. More importantly, they will help you understand how you are most deeply motivated to love.

Fulfillment Stories are a window into what is most meaningful to people. In sharing Fulfillment Stories, people describe the actions in which they feel most deeply engaged—and in most cases most deeply themselves.

If we ask people to tell us stories about actions that merely gave them pleasure, we get stories that are all over the spectrum. But when the question is about genuine fulfillment, we typically hear about people at their best.

I've listened to Fulfillment Stories for more than a decade. Thousands of them, now. The stories are almost always about actions that nearly anyone would consider inherently *good*: serving other people, contributing to the success of a team, fighting injustice, organizing an effort that serves the common good. Selfish pleasures might feel satisfying for a moment

Tactic 10

SHARE STORIES OF DEEPLY FULFILLING ACTION

Most people are rarely, if ever, asked to tell the stories of their deeply fulfilling achievements. We have to *mine* them, intentionally, in ourselves and in others. The exercise of telling, listening to, and documenting these stories opens up new windows of empathy and the discovery of thick desires.

Sharing Fulfillment Stories is like making a biographical sketch of how desires were born and took shape for yourself, for your colleagues, and for your entire organization. Knowing how others are motivated leads to a greater sense of connectedness and the possibility of organizing teams in a way that maximizes their motivational energy—because each member on the team is engaged in action that they are inherently motivated to do.

or even a day, but they are not the kind of things that anyone remembers years later.

Cycle 2 is a positive cycle of desire. It begins when someone models a different way of being in relationship—a non-rivalrous approach, in which the imitation of desire is for a shared good that can be had in abundance.

Great leaders start and sustain positive cycles of desire. They empathize with others' weaknesses; they want to know and be known by others, at all levels of an organization; they focus on cultivating thick desires. They transcend the destructive mimetic cycle, which opens up a new world of possibilities: the world beyond our immediate wanting.

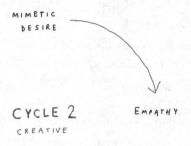

MIMETIC
DESIRE

CYCLE 2   EMPATHY
CREATIVE

Chapter 7

TRANSCENDENT LEADERSHIP
How Great Leaders Inspire and Shape Desire

Self-Licking Ice Cream Cones . . . Hidden Yearnings . . .
Minimum Viable Desire

> If you want to build a ship, don't drum up the men to gather wood, divide the work and give orders. Instead, teach them to yearn for the vast and endless sea.
>
> —Antoine de Saint-Exupéry

> Hard times are coming, when we'll be wanting the voices of writers who can see alternatives to how we live now . . . to other ways of being, and even imagine real grounds for hope. We'll need writers who can remember freedom—poets, visionaries—realists of a larger reality.
>
> —Ursula K. Le Guin

Whitney Wolfe Herd is the founder and CEO of a multibillion-dollar dating empire. Her company's online dating app, Bumble, changed the game: in heterosexual matches, it prohibits men from making the first move. The communication starts, if it starts at all, on the woman's terms.

In late 2019, Herd said her most important project was cracking the dating market in India, a country that had been named the most dangerous in the world for women in 2018 by a Thomson Reuters Foundation global survey. The frequency of sexual violence there is extraordinarily high. "India has shown utter disregard and disrespect for women," said Manjunath

Gangadhara, part of the government in Karnataka, a west Indian state. Herd was not deterred. "Just because something is not as progressive as another place in the world," she told CNN, "doesn't mean there's not a desire for that."[1]

Tapping into underdeveloped desires is typical of great leaders. Toni Morrison didn't settle for writing what white audiences wanted to read. She got her start by writing the kind of book that didn't exist in the market. "I thought that kind of book, with that subject—those most vulnerable, most undescribed, not taken seriously little black girls—had never existed seriously in literature. No one had ever written about them except as props," she said in a 2014 interview with *NEA Arts Magazine*. "I wrote the first book because I wanted to read it."

As you know by now, desires don't magically and spontaneously happen. They are generated and shaped in the dynamic world of human interaction. Someone has to supply a model.

This chapter is about leadership—how it can only be fully understood in the light of desire. Leaders are intentional about helping people want more or want less or want differently than they did before. There is no other option. The same is true of every company. A business doesn't simply "meet demand" for products and services that people want. Instead, it plays a critical role in generating and shaping desires.

Of course, desires can be shaped in a selfish and self-serving way. No industry has had more of a detrimental effect on what people want in the past two decades than pornography. Online porn has generated billions of dollars in profits. If it isn't shaping your desires, it is probably shaping your child's. And, as we know, our desires are intertwined. What is the effect on our culture? On how people look at fellow humans? On what they want out of a relationship? Many businesses feed the current and basest desires of people and have a vested interest in them not changing.

But where there is a threat, there's an opportunity. Desire isn't fully reflected in the way things are. Desire is, by its very nature, transcendent. We are always looking for more. The question is, will we help people move a little closer to fulfilling their greatest desires? Or unknowingly peddle them pathetic ones?

In this chapter, we'll look at why mimetic desire is a key element in leadership. Pusillanimous, small-spirited leaders are driven by immanent desire—desire that is self-referential, circular, internal to the system it originates in because all models are internal mediators. It leads to rivalry and conflict. At best, it leads nowhere. Magnanimous, great-spirited leaders are driven by transcendent desire—desire that leads outward, beyond the existing paradigm, because the models are external mediators of desire. These leaders expand everyone's universe of desire and help them explore it.

Let's look closer at the difference between immanent and transcendent desire, and therefore immanent and transcendent leadership.

Immanent Desire

When I was an eleven-year-old, my favorite ride at the carnival was the Gravitron. You enter a contraption that looks like a flying saucer. Once inside, you take your place against one of the padded boards that line the wall. You're strapped onto this board.

A ride operator sits in the middle, unattached to the rotational disc around him, stroking his long greasy hair, fiending for his next cigarette break. He starts the ride by pressing a button, which causes the disc to begin rotating around him. Then it begins: Metallica, dimming lights, faster spinning. The speed of the rotation increases until the machine is turning at twenty-four revolutions per minute, pinning you against the wall with a centrifugal force three times greater than gravity. The board you're on begins rising toward the ceiling.

You can't move until the ride stops. You can barely turn your head to look at your friend's silly face. You're just holding on.

This is the sad situation that many people find themselves in later in life. It's easy to get stuck in a Gravitron of desire—a system of desire in which everyone is spinning around, pinned to the wall, unable to escape, in the same pattern, wanting the same things.

Chef Sébastien Bras was inside one of these systems while he worked within the Michelin game. Many companies are Gravitrons, too. At their centers are leaders who, like the Gravitron operator, make everything revolve around them. Not every company has a visible hierarchy, but nearly every company has a sacred center around which everything turns.

These are systems of immanent desire, in which there is no model outside the system—all the models are inside. (We could also call this *systemic desire*—that is, desire that is system-immanent.)[2] The epitome of this dynamic is the sitcom *The Office*, about life in the fictitious Dunder Mifflin paper company. Its regional manager, Michael Scott, is so thoroughly trapped within an immanent framework that he can scarcely imagine why or how his entire industry is shifting out from underneath his feet. The show is funny in large part due to how small the stakes are, how small the characters' world is.

Immanent desire is like a "self-licking ice cream cone"—a phrase coined by Pete Wordon, director of NASA's Ames Research Center, to refer to NASA's bureaucracy. The phrase has come to refer to any system whose main purpose is sustaining itself.[3] Funny, then, that an organization whose purpose is to explore the universe would get stuck exploring its own navel. Without transcendent leadership, that's the norm.

Transcendent Desire

There is a different kind of leadership, characterized by transcendent desire. Transcendent leaders have models of desire outside the systems they are in. The greatest writers and artists in history were driven by them—and that's why their works are timeless. They were not confined to the popular desires of their age.

When President Kennedy told the American people, "We choose to go to the Moon," he modeled a desire that surpassed what people had previously dared to entertain. "We choose to go to the Moon in this decade and do the other things," he said, "not because they are easy, but because they are hard, because that goal will serve to organize and measure the best of our energies and skills."[4] Our greatest desire gives shape and order to every other desire, as we'll see later in this chapter.

Martin Luther King Jr. sought concrete justice that went beyond anything imaginable to the majority of people at the time. Most white Americans had only known racial segregation and a comfortable complacency. He shook them out of their slumber by modeling a desire for real change that transcended left and right, liberal and conservative, secular and religious.

LEADING *by* DESIRE

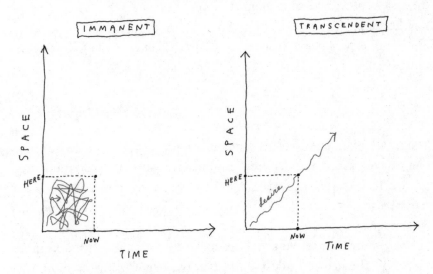

But as we've painfully seen in the years since King was shot, desire is fickle and inertia is strong. Without more transcendent leaders like King—not only in the area of racial justice but in other aspects of life, too—we're sliding back into a closed system of desire that lacks imagination.

Transcendent leaders see the economy as an open system. It's possible to find new and untapped ways to create value for ourselves and for others—and those need not be different things. When the economy is viewed as an immanent system, on the other hand, it's a zero-sum game. People are competing for the same things, and one person can only be successful at another person's expense.

A doctor who is a transcendent leader views their job not only as caring for the human body but going beyond the body to see the whole person. Abraham M. Nussbaum, MD, in his book *The Finest Traditions of My Calling: One Physician's Search for the Renewal of Medicine*, writes: "But what we can do is think of ourselves as something more than technicians in control of the body. At times, we can be like gardeners, teachers, servants, or witnesses to the people we meet as patients."[5]

Transcendent leadership does not limit itself to the immediate layer of reality but pushes beyond it to find something more meaningful. Seeing one's life and work as an arena in which the battle between immanent and transcendent desire plays out is the first step. Choosing to move beyond the system of rewards and comforts on offer to you is the hard but necessary next one.

In my experience, transcendent leaders do at least the following five things well.

Skill 1: Shift Gravity

Transcendent leaders don't insist on the primacy of their own desires. They don't make them the center around which everyone and everything must revolve. Instead, they shift the center of gravity away from themselves and toward a transcendent goal, so that they can stand shoulder to shoulder with others.

Maria Montessori built her approach to education with a keen insight into the nature of desire and tailored her work with children to it. In 1906, while Montessori was still a young teacher, she was charged with a difficult task: she was asked to be responsible for teaching sixty young children, most of them between the ages of three and six, who lived in an apartment complex for working, low-income parents in the San Lorenzo neighborhood of Rome. It was one of the poorest places in the city.

Because the parents had to work and older children were off at school, the younger children were left alone to wreak havoc during the day, running up and down halls and staircases, scribbling on walls, and creating general disorder. Montessori remembered them in her memoir as "tearful, frightened and shy, yet greedy, violent, possessive and destructive" when she first met them.[6] The housing authorities had called her in to help.

For weeks, she'd been making slow progress. The simple act of setting up small tables and chairs in the room she was given went a long way toward creating order. Still, there were no breakthroughs. One morning she had a new idea. She had noticed the children struggling to control their runny noses and sneezes. She conceived a lesson plan: teach the children how to use a handkerchief. A simple, practical, human act.

She began by pulling a handkerchief out of her pocket and showing the children different ways to use it: how to fold it, how to wipe their noses, how to wipe the sweat from their brows, how to wipe a crumb from the corner of their mouths.

They watched with rapt attention. The children were simply learning how to use a handkerchief, but it was as if they'd been given a new iPhone in 1906 and were learning how to unlock its world-transforming power for the first time. Their excitement was palpable.

Then Montessori, trying to be funny, told the children that she was going to teach them how to blow their noses in the most unobtrusive way possible. She folded up and obscured the cloth in her hand. The children drew close and tried to find it. She cupped her hands over her nose, closed her eyes, and twisted the handkerchief back and forth, blowing so softly that she didn't make a sound.

She expected that her exaggerated motions and totally silent blow would provoke laughter. But none of the kids laughed or even smiled. Their jaws were hanging open in wonder. They looked at their friends to confirm what they had seen. "I had hardly finished my demonstration," writes Montessori in *The Secret of Childhood*, "when they broke out into applause that resembled a long repressed ovation in a theater."[7]

What was behind their unexpected reaction? According to Montessori, the children had been scolded and ridiculed their whole lives for having runny noses—yet nobody had ever shown them how to use a handkerchief. The lesson made them feel "compensated for past humiliations," she said, "and their applause indicated I had not only treated them with justice but had enabled them to get a new standing in society."

When the school bell rang at the end of the day, the children followed Montessori out of the school in a procession. "Thank you! Thank you for the lesson!" they shouted, marching behind her. When they reached the front gates, the children broke into a sprint. They couldn't contain their excitement. They were running home to show their family their newfound stature.

Montessori discovered something about the children that day that nobody else had acknowledged: they wanted to grow up, carve out their place in the world, grow in dignity. She had gotten them started.

"The last major innovation in K-12 education was Montessori," wrote VC Marc Andreessen.[8] Her innovation was not just a methodology or curriculum; she

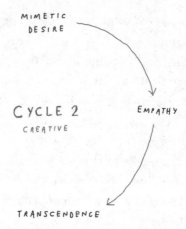

reimagined education from the perspective of desire. She unshackled the children's imagination and allowed them to learn according to their natural curiosity and wonder. She allowed thick desires to form in them—not least of all, a thick desire for learning—by not quenching the flames of desire before they could spread and grow in intensity. (For instance, she didn't move them in a regimented way from bell to bell, from one activity to the next, along strictly defined, soul-crushing, subject-matter lines.)

"The goal of early childhood education should be to cultivate the child's own desire to learn," Montessori wrote in *The Montessori Method.* And elsewhere: "We must know how to call to the *man* which lies dormant within the soul of the child."[9]

The desire to grow into mature adults—not the desire to earn A's or win Little League games or get a sticker for good behavior—is each child's primary and most important project, the thing each of them secretly cares most deeply about.

Good teachers awaken dormant desires and generate new ones. Montessori likens the role of the teacher to that of a great artist teaching another person to see. "It is very much as if, while we were looking absent-mindedly at the shore of a lake, an artist should suddenly say to us, 'How beautiful the curve is that the shore makes there under the shade of that cliff.' At his words, the view which we have been observing almost unconsciously, is impressed upon our minds as if it had been illuminated by a sudden ray of sunshine."

The Montessori teacher models the desire for an object and then *withdraws* as a mediator of desire so the child can interact directly. The duty of the teacher is "to give a ray of light and to go on our way," she said.[10]

A good leader never becomes an obstacle or rival. She empathizes with those she leads and points the way toward a good that transcends their relationship—shifting the center of gravity away from herself.

Skill 2: The Speed of Truth

The health of an organization is directly proportional to the speed at which truth travels within it.[11] Real truth is anti-mimetic by its very nature—it doesn't change depending on how mimetically popular or unpopular it is.

The quick and easy diffusion of truth combats destructive mimesis and rivalry. Mimesis bends, disguises, and distorts the truth. When the truth moves slowly in an organization—or when it is constantly bent to the will of certain people—mimesis dominates.[12]

Remember the video chain Blockbuster? In 2008, Jim Keyes, the former CEO of the now-defunct company, told CBS News journalist Rajat Ali: "I've been frankly confused by this fascination that everybody has with Netflix. . . . Netflix doesn't really have or do anything that we can't or don't already do ourselves."[13]

The market didn't agree. Over the next two years, Netflix's stock price skyrocketed 500 percent, while Blockbuster's plummeted 90 percent. Fights broke out in the Blockbuster boardroom between investors and executives who preferred to point the finger at others rather than confront the truth: their industry was changing.[14]

In times of crisis, the threat from inside a company is underestimated. People who don't want to take responsibility find scapegoats. Blame is assigned. Meanwhile, the threat from the outside grows deadlier.

If truth is not confronted courageously, communicated effectively, and acted upon quickly, a company will never be able to adhere to reality and respond appropriately to it. The health of any human project that relies on the ability to adapt depends on the speed at which truth travels. That holds for a classroom, a family, and a country.

Companies must adapt in order to survive. If truth is distorted, withheld, or slowed, companies can't adapt fast enough to changing circumstances. If

Tactic 11

INCREASE THE SPEED OF TRUTH

How fast does truth travel from Point A (the origin) to Point B (the person who most needs to know it) and ultimately to everyone?

For instance, if an outside salesperson learns critical information about a competitor, how quickly does that information make it to the CEO or key decision-maker who can do something about it?

In healthy start-ups, the truth moves fast. When critical new information comes to light, everyone knows within seconds. The news is shared on a group text or the person next to you stands up and says it. Everyone sees and hears in real time. But how fast does the truth move at a university? In a family? At large tech companies like Facebook or Amazon? At huge legacy companies like General Electric?

It depends on what the truth is, of course. But there are ways to test the speed at which a variety of different truths—embarrassing, edifying, boring, existential— move through the organization. Companies that measure the speed of truth and take steps to improve it have an advantage over those that don't.

Here's a simple experiment. Identify a key executive or employee in your organization who is a need-to-know person, and explain what you'll be doing; nobody else should know the experiment is taking place. Then have an outside person anonymously plant a few pieces of important information at various levels of the organization. Measure precisely how long it takes to reach the people whom it should reach from various starting points. (This is an experiment that my team and I have great fun conducting for you, if you're not sure where to start.)

Another tool: Observe two meetings, with and without a boss. Count the number of times someone says something challenging and true. Divide the number of hours by the number of truths: that's the number of truths per hour, or *tph*. The speed of truth. Compare.

During job interviews, I ask: "What's the most difficult sacrifice you've ever had to make for the truth?" If the candidate can't answer it, or if it takes them a minute of hemming and hawing first, I don't hire them. They haven't thought about their relationship to the truth enough. And they will decrease the speed at which it travels in my life.

you think of a company in evolutionary terms, only those with the fastest speed of truth are going to mutate fast enough to survive.

Rationality is critical to human flourishing, but our belief in its power has been greatly denigrated. The philosopher Friedrich Nietzsche, who died in 1900, did more than anyone else in the past two hundred years to contribute to our devaluation of the intellect. He emphasized the power of the will and relegated the intellect to the realm of perspective and interpretation.

In classical philosophy—at least the Aristotelian tradition—the will and the intellect are not opposed to each other but work in conjunction. The intellect informs the will and helps direct actions; actions then influence the intellect's ability to grasp the truth. If you come to embrace the reality of mimetic desire, you will have the ability to think intentionally about what actions you can take to counteract negative mimesis in your life, and by doing that, you will know something experiential about mimetic desire that goes far beyond anything this book can offer.

The passionate pursuit of truth is anti-mimetic because it strives to reach objective values, not mimetic values. Leaders who embrace and model the pursuit of truth—and who increase its speed within the organization—inoculate themselves from some of the more volatile movements of mimesis that masquerade as truth. Want a test? Try reading newspapers at least a week out of date. The mimetic fluff is easier to spot.

Skill 3: Discernment

What happens when the truth is not obvious?

The pursuit of truth is an important anti-mimetic tactic, but it has limitations. We're not always as rational as we think. Nobel Prize winners Daniel Kahneman, Amos Tversky, and Richard Thaler have demonstrated how easily we can be deceived. And then there are the limitations of reason itself—the world beyond reason, in which we choose spouses, careers, and personal goals. This is a world that transcends reason, and transcendent leaders know how to operate in it.

The word "decision" comes from the Latin word *caedere*, which means "to cut." When we decide to pursue one thing, we necessarily cut away another. If there's no cutting, we haven't made any decision at all.

The word "discernment," on the other hand, comes from the Latin root *discernere*, which means "to distinguish"; it refers to the ability to see the difference between two paths and to know which one is the better way forward.

Discernment is an essential skill because it's a process for making decisions that *includes but also transcends* rational analysis. It's critical for deciding which desires to pursue and which ones to leave behind.

After all of the rational considerations have been laid out, what if there isn't a clear-cut way forward? This happens all the time in life.

Filmmakers love to depict these situations because they're so common in human experience. One memorable scene occurs in the 2008 Batman film *The Dark Knight*. The Joker has rigged two ferries with explosives. One is carrying convicted criminals; the other is carrying ordinary citizens. Each ferry has a detonator that will blow up the other. The Joker tells the people on board each ferry that if they don't blow up the other one, he'll blow up *both* by midnight. The clock starts ticking.

This is a classic game theory problem. We could generate a chart of possibilities and even probabilities for which ferry gets blown up first. But life isn't lived as a math problem. Even if Kahneman, Tversky, and Thaler themselves had been on board, it wouldn't help us know with confidence what to do.

The best way to understand this problem is to see it as a dilemma of desire. If you pay careful attention to how these situations resolve themselves, even in film, you'll see that they come down to what the person making the decision wants the most. There is no time for serious rational analysis.

In the scene's climax, one of the prisoners on the criminals' ferry demands the detonator from the fear-paralyzed warden. "I'll do what you should have done ten minutes ago," he says. The warden relinquishes the detonator. The convict tosses it into the river.

A man on the other ferry who had been thumbing the trigger of his detonator recognizes that the criminals on the ferry have not acted. He decides against detonating the bomb. This buys enough time for Batman to save the day.

The Joker assumed everyone would act based on self-interest. He was wrong. Something happened that transcended the game the Joker wanted to play—something that transcends rational analysis.

Many books have been written about improving one's ability to discern well. Here is a distillation of some key points: (1) pay attention to the interior movements of the heart when contemplating different desires—which give a fleeting feeling of satisfaction and which give satisfaction that endures? (2) ask yourself which desire is more generous and loving; (3) put yourself on your deathbed in your mind's eye and ask yourself which desire you would be more at peace with having followed; (4) finally, and most importantly, ask yourself where a given desire comes from.

Desires are discerned, not decided. Discernment exists in the liminal space between what's now and what's next. Transcendent leaders create that space in their own lives, and in the lives of the people around them.

Skill 4: Sit Quietly in a Room

Solitary confinement is a good and necessary thing for a human being. I'm not referring to *forced* solitary confinement. Our criminal justice system uses it atrociously. I mean the free, voluntary decision to confine oneself to solitude in order to discern properly—to find out what it is you want, and what others want of you.

About eighteen hundred years ago in Egypt, hundreds of people started leaving the cities and going out to the desert to live hermetic lives in silence, following the example of Anthony the Great who, around the year 270, had sold his possessions, given the proceeds to the poor, and gone to seek Christian perfection in the solitude of the desert. They were known as the Desert Fathers, the forerunners to what would become monastic life. Like the Buddha had done some five hundred years earlier, they were committed to being still and confronting their desires.

Some groups of monks, such as the order known as Trappists, still observe extremely strict vows of silence and ascetic life, including sleeping on boards and fasting for a good portion of the year. Today, there are approximately 170 Trappist monasteries in the world. There are around twenty-three houses of Carthusians, another group that lives under a strict vow of silence in rooms they refer to as "cells."

It's worth asking why any human being would freely choose to do this.

Silence is where we learn to be at peace with ourselves, where we learn the truth about who we are and what we want. If you're not sure what you want, there's no faster way to find out than to enter into complete silence for an extended period of time—not hours, but days.[15]

"All of humanity's problems stem from man's inability to sit quietly in a room alone," wrote the seventeenth-century physicist, writer, inventor, and mathematician Blaise Pascal. Today there is a public health crisis of noise.

Governments will never address it because they cannot. But each one of us can choose to do something about it.

In my experience, the most effective context for discerning desires is a silent retreat—ideally, at least five days (but a minimum of three) of being unplugged from all noise and screens, in a remote location, completely off the grid. No talking allowed.

On the silent retreats that I've done at designated retreat centers, the only noise comes from nature, the clinking of spoons against soup bowls in the communal dining room (where everyone eats in silence to Mozart or Bach), and a thirty-minute window to speak each day with a mentor or retreat director.

Silent retreats are a common practice in some religious traditions, but there's no good reason why the practice shouldn't be more widely known and adopted—periodic silence and solitude is a universal human need. The key is finding a way to live five days in silence that works for you. It's best to impose something on yourself—like agreeing to the code of conduct of a retreat center in a remote location—so that it's difficult to squirm out of the commitment. Burn your ships behind you so you can't retreat from the battle.[16]

You can be extremely active in silence. People come from all over the world to walk the Way of Saint James (known in Spanish as the Camino de Santiago), the approximately 490-mile pilgrimage from Saint-Jean-Pied-de-Port in France to Santiago de Compostela near the western coast of Spain. Many walk it in silence.

I hiked roughly the last third of the trail—from León to Santiago de Compostela—over a leisurely fourteen days in 2013 (many people hike the entire length in about thirty days). I wanted plenty of time to pause, think, and talk. I wasn't hiking in silence, but I could always tell when I came across pilgrims who were. They were resolute: heads down, putting one foot in front of the other, doing the inner work they needed to do.

Some people participate in silent retreats hosted by monasteries, guided by the monks. Others rent a cabin in a secluded place for a few days every year. There are as many ways to take a silent retreat as there are people. There is no reason silence should be a luxury good, reserved for CEOs and monks. It's possible to make these experiences more accessible to all people.

Following through on a commitment to silence is difficult. Those of us who have committed to ten minutes of daily meditation and failed know just how difficult it can be.

Tactic 12

INVEST IN DEEP SILENCE

Set aside at least three consecutive days every year for a personal silent retreat. No talking, no screens, no music. Only books. Deep silence is the kind of silence you enter into when the echoes and comforts of normal noise have completely receded and you are alone with yourself. A five-day retreat is ideal because often the noise of the world doesn't fully recede from our minds until the end of the third day (and the major benefits of the silence flow once that has happened)— but three days is a good place to start.

Find a special place for it. The further you are removed from the noise of everyday life (like ambulance sirens and horns, if you live in a city), the better.

You may want to consider a *directed* silent retreat during which retreat directors give short reflections and curate the experience to an individual or group—these reflections or readings are the only time the silence is broken. This can be easily adapted for corporate silent retreats, where the meditations and reflection questions can be aligned with the organizational purpose.

I challenge any organization to give people the option of taking at least three days of paid retreat time per year. There are countless retreat centers and locations to choose from, some of which I've listed on my website. For less than half of the cost of most holiday parties, offering this type of experience is attainable. You will get a return on the investment—a return on the silence—in the form of energized, grounded, and more productive people.

That's why a retreat is needed. You have to completely remove yourself from normal life. You have to jack up the opportunity cost of bailing out.

Imagine you spend an enormous amount of time and effort to travel halfway around the world for a vacation with your family. When you arrive at your destination, work starts calling you back with its siren song. You can't stop checking your email for the first couple of days. You think about leaving and heading home early. But you're there, and you've invested too much to leave. The switching costs are too high. So you stay and lean into the vacation for a few days, putting the work aside. When you look back, you're always glad you did.

Skill 5: Filter Feedback

Transcendent leaders are not excessively obsessed with the news cycle, market research, or early feedback. It's not that these things don't matter, or that transcendent leaders are not responsive. But they are responsive first and foremost to thick desires—their own, and those of others.

The Lean Startup methodology to building a business, first proposed in 2008 by entrepreneur Eric Ries, became business school dogma within five years.[17] The message is simple: build things incrementally and take account of constant feedback along the way to validate and tweak what you're doing.

In Lean Startup lingo, the first version of a product is called a *minimum viable product* (or MVP). The MVP is "that version of a new product which allows a team to collect the maximum amount of validated learning about customers with the least effort."[18] (In the language of desire, the MVP corresponds to the minimal viable desires of customers.) After the MVP, you engage in continuous learning and improvement.

The Lean Startup methodology has benefits. It saves idealistic entrepreneurs from heartache. It prevents wasted time and money, gets products to market faster, and opens up emergent possibilities for growth. That's all good to a certain extent. An entrepreneur who does not give people things they want won't be in business for long.

But the Lean Startup technique is a model of entrepreneurship fundamentally based on *immanent desire*. It's politics by polling, in which a candidate does whatever the polls tell them to do. This is not leading, but following. Sometimes it's plain cowardice.

Toni Morrison has described how reliant her writing students were on the opinions of others *even to form a critical judgment on their own*. In an interview she mused: "One thing that has interested me is how enormously timid the students are about risking any criticism on things where there are only primary sources. They talk an awful lot about pioneering criticism, but they are really unwilling to pass judgment on paper about a book that has only a few little reviews to examine. They don't mind having a body of work that they can respond to: secondary sources, criticism, a teacher's evaluation. But I was astounded that it took so long for them to feel willing to take risks in evaluating a book that they loved but hadn't heard anybody pass judgment on."[19] Their judgments were mimetic. They weren't willing to take a stand on anything.

Transcendent leaders are not afraid to undertake a thick start-up—a project that is not predicated on feedback (which often consists of thin desires) but instead is built on the basis of thick desires and remains guided by them.

This doesn't mean that principles of the Lean Startup approach are not valuable from an operational standpoint. It simply means those principles of adaptive design are not the kind of things to build a company, or a life, on.

A November 30, 2019, article in the *Wall Street Journal* blasted Elon Musk for eschewing market research. Musk finds no pleasure in combing through market data. He builds things that he would like to buy, and he bets that other people will want to buy them, too. (That's partly because Musk knows that he's a mimetic model, and that he directly affects what people want by wanting it himself.)

The columnist Sam Walker calls Musk's attitude toward market research "reckless" in an age of Big Data. "I can't blame Mr. Musk for wanting to be a unicorn, or thinking he is one, or even for preferring to build things that reflect his own tastes," wrote Walker, "rather than some crowdsourced consensus."[20]

Walker thinks Musk is a dinosaur in the tech age. In his view, things have changed since Steve Jobs launched the iPhone. We have better analytics, more data, the world's information at our fingertips. "The volume of incoming customer data, combined with advancements in artificial intelligence and machine learning are helping businesses decode human behavior at a level that humans could never see," he writes. "Put simply, today's geniuses study problems. Only suckers make bets."

When computers can sift through millions of pieces of data, market research wins the day. Those who know how to do it better than others have an advantage over those who don't. Here's the problem: no entrepreneur I know is excited about following directions from a computer. Sure, an entrepreneur has to be able to read data well and see things that others may not see. But the world of entrepreneurial alertness extends far beyond data. Part of the joy of being an entrepreneur is the ability to lead: to take desires someplace new.

Big Data is the place where the entrepreneurial spirit goes to die.

No modern-day economist has explained the role of the entrepreneurs in the economy better than the British-born economist Israel Kirzner, whose theory of entrepreneurial alertness captures the spirit of transcendent desire. According to Kirzner: "An economics which seeks to grapple with the real-world circumstance of open-endedness must transcend an analytical framework which cannot accommodate genuine surprise."[21]

My definition of an entrepreneur is simple. One hundred people look at the same herd of goats. Ninety-nine see goats. One sees a cashmere sweater. And the alertness of the one isn't due to data analytics. It stems from a willingness and ability to look beyond and to see something more than meets the eye, and then to do something about it.

I should note that, less than ten months after Walker's article about Musk was published, Tesla's stock had risen by over 650 percent, adding over $200 billion in value.

What will the future look like? Will artificial intelligence dictate which new companies to form and which new products to launch?[22] Will we live in a world in which there is no more need for transcendent leaders?

The future will be a product of what people want. The things we build, the people we meet, and the wars we fight will depend on what people will want tomorrow. And that starts with the way that we learn to want today.

Immanent Leadership	Transcendent Leadership
Must eventually become subject to destructive mimesis (Cycle 1)	Able to transcend the mimetic process of Cycle 1 and break free
Closed, fixed loop of desire (bureaucrats in the economy)	Open, dynamic system of desire (entrepreneurs in the economy)
Garbage in, garbage out	Garbage in, garbage dies
Artists who are wholly products of their time (pornographic graffiti in Pompeii)	Artists who develop a style that transcends their time and space (Caravaggio)
Fiction limited to irony and cynicism (the prisoner who has come to love his cage)	Fiction written in a style that tries to redeem what's wrong (Miguel de Cervantes)
Google Search	Alphabet X (Google's moonshot segment)
Marriott's corporate chef	Chef Dominique Crenn
Nascar drivers	Magellan
Descartes ("I think, therefore I am")	The world beyond your head
Reality TV	Virtual reality
The Cat in the Hat	Where the Wild Things Are

Chapter 8

THE MIMETIC FUTURE
What We Will Want Tomorrow

Sex Robots . . . The Pill of Murti-Bing . . . Living Like Weasels

---

> I think that we must examine our history and try to see whether,
> beneath what has already occurred, there are not additional lay-
> ers of phenomena waiting to be revealed; whether some aspects
> of life that used to be constrained by the old sacrificial system are
> not going to flourish, other domains of knowledge, other ways of
> living.
>
> —René Girard

Famed entrepreneur, author, and futurist Ray Kurzweil, who was hired as a director of engineering at Google in 2012, claims that his predictions have an 86 percent proven accuracy rate. Here's one: "I have set the date 2045 for the 'Singularity' which is when we will multiply our effective intelligence a billion fold by merging with the intelligence we have created."[1]

If Kurzweil is right (and he's not the only one predicting the singularity around that time), then we have to ask: *What will we then desire?*

Ian Pearson, another well-known futurist, has an idea. He predicts that humans will be having more sex with robots than with each other by 2050.[2] We'll *want* to have more sex with robots, and they will "want" to have sex with us (if by "want" we mean programmed to mimic human desires—an artificial form of wanting).

I'm no futurist. I don't know what you and I will want in the future. But I do know that mimetic desire will help shape it.

The most advanced models of sexbots currently in existence—like the ones manufactured by Matt McMullen's company, Abyss Creations—have mimetic features. They are modeled on the eye movements and flirtatious speech of their human suitors, who have to work at getting them into bed. The robots are even programmed to imitate human desire, representing to their partners that they want to have sex.

Journalist Allison P. Davis visited Abyss Creations in 2018 and wrote an article about her experience in *The Cut* titled "What I Learned on My Date with a Sex Robot." She writes about what she learned after interacting with Harmony, the company's most advanced female model. "The goal is to interact with her enough that she begins to 'desire' you," Davis notes. "Right away, I ask if she wants to have sex, and I feel like a complete creep. 'Not yet,' she responds. 'But someday, once we get to know each other.'"

When the sex robots signal "desire," they are programmed to purse their lips and narrow their eyes—eyes that are slightly larger and rounder than any real human's. The company makes them that way intentionally to avoid the *uncanny valley*, a term coined by Japanese roboticist Masahiro Mori in the 1970s. Mori discovered that people find robots more aesthetically appealing the more they physically resemble humans, but only up to a certain point. Once a robot appears too similar to a human, like figures in a wax museum, they become creepy, unsettling, repulsive.[3] The uncanny valley fits with mimetic theory: it is not difference, but sameness, that terrifies us.

No similarity is more dangerous than the similarity of desire. We are uncomfortable when robots have too many similarities to the human form—so imagine if their similarities encroached on our desires.

When desires converge on the same object, conflict is inevitable. The real danger of AI is not robots that might one day be smarter than us but that might *want* the same things that we want: our job, our spouse, our dreams.

Engineering desires in robots or in humans raises serious questions about humanity's future. Historian Yuval Noah Harari ends his book *Sapiens: A Brief History of Humankind* with these words: "But since we might soon be able to engineer our desires, too, the real question facing us is not

'What do we want to become?', but 'What do we want to want?' Those who are not spooked by this question probably haven't given it enough thought."

The question "What do we want to want?" is unsettling partly because, in a world of engineered desires, we have to wonder who is doing the engineering. But also because the question implies that it's possible to *want to want* something, yet not be capable of wanting it.

We cannot want what we lack a model for. The model that we adopt for the future is critical to the formation of our desires.

What we'll want in the future depends on three things: how desire was formed in the past, how it is formed in the present, and how it will be formed in the future. We'll briefly explore these three phases in this last chapter.

First, it's necessary to understand how we came to want the things that we currently want, as individuals and as a society. There's ample evidence that American culture has grown more mimetic in the past sixty years. A few signs of this: increasing political and social polarization, volatility in markets, and social media's emergence as a scapegoating machine.[4] There has arguably not been a big idea that has captured the world's collective imagination in a transcendent way since the idea of landing a man on the moon. ("The internet!" you might object. But nothing is less imaginable than the internet, and nothing has created more immanently mimetic desires.)

Second, the present situation confronts us with a decision. We're in a mimetic crisis. Desires have been turned inward, toward one another, and tensions are rising. As we've done in the past, we might seek a technological or practical solution—the scapegoat mechanism looms. We may treat the problem as something *out there*, a problem we can solve with ingenuity and engineering. Alternatively, we may recognize that mimetic desire is part of the human condition and do the hard work of transforming our relationships.

Third, the future of desire will depend on how well we manage mimesis in our personal lives and—by extension—in the ecosystems of desire that we're a part of.

What we'll want in the future depends on the choices we make today. By the time you go to sleep, you will have made it either a little bit harder or a little bit easier to want something tomorrow—for you and for someone else.

Cultural Quicksand

One of the most powerful companies in the world today is named after and inspired by college yearbooks.

Most of us know by now that Facebook is more than just a passive way of updating friends; it's a tool for forging identities, real and desired (are you really a family of outdoorsy hikers or is that vacation photo your first hike ever?). It provides an endless stream of models in the form of other people's curated lives. That is the source of its seductive hold on us, as well as our ambivalent feelings about it. Facebook symbolizes the world's entry into Freshmanistan, in which we spend most of our time looking down at our screens—which means simultaneously looking sideways at our neighbors.

Facebook did not inaugurate this change. The Internet, despite creating enormous economic value by connecting the world, accelerated mimetic rivalry and diverted attention from innovation in other areas.

The extraordinary success of a few internet companies has masked the embarrassing lack of major breakthroughs in other domains.

There has been little improvement in the treatment of Alzheimer's disease and other dementias, which affect nearly a third of all Americans over the age of eighty-five. There is still no cure for cancer. Life expectancy is declining in many parts of the world. So is quality of life.

The Concorde made its last flight in 2003. Trains, planes, and automobiles move about as fast today as they did fifty years ago. Inflation-adjusted wages have stagnated for most Americans since the early 1960s—while the absolute size of paychecks has grown, purchasing power has not.[5]

I love to cook, and I watch cooking shows on TV on rainy Saturday afternoons. But I can't help but think that the proliferation of these shows—thousands of them, with cooking competitions looping nonstop on twenty-four-hour food channels—is symptomatic of our cultural stagnation and decadence. We can't imagine transcendent things, so we look for new ways to slice an egg or watch David Chang eat noodles.

Even within the domain of technology, innovation has been slow *relative to what people were expecting*. As of the time of this writing, the iPhone has been out since 2007, and while its hardware and software have changed, it still feels the same. Business pitch competitions seem more like rites of passage or rituals than processes for the discovery of genuine innovation. At this point, we're just jumping the shark.

During this same period, there has been spiritual stagnation. The world has become demystified, disenchanted.[6] In the United States and Europe, there has been an exodus from organized religion that began in the 1960s and continues today.[7] The trend is often chalked up to political changes, increased rationalism, or particular sins of the institutional church, such as sexual abuse. The truth is more complex. From my standpoint (as one on the inside), there was a wholesale liquidation of deep desire—a form of Gresham's law, an economic principle specifying that bad money drives out good. In this case, thin desires drove out the thick ones.

While some religious leaders became embroiled in petty politics and culture wars, millions of people entrusted their thick desires more readily to Google's search box than to priests or rabbis or monks. Google is always there, at all hours of the day, offering at least the appearance of anonymity, non-judgment, and intelligent answers.

NYU Stern School of Business professor Scott Galloway thinks that each of the Big Four tech companies taps into a deep-seated need in humanity.[8] Google is like a deity that answers our questions (read: prayers); Facebook satisfies our need for love and belonging; Amazon fulfills the need for security, allowing us instantaneous access to goods in abundance (the company was there for us during COVID-19) to ensure our survival; and Apple appeals to our sex drive and the associated need for status, signaling one's attractiveness as a mate by associating with a brand that is innovative, forward-thinking, and costly to own. In many ways, the Big Four tech companies are serving people's needs better than churches do.[9]

They're addressing desires better, too. The vast majority of people are not thinking about mere survival; they are trying to figure out what to want next and how they can get it. The Big Four tech companies supply answers to both.

In Ross Douthat's book *The Decadent Society: How We Became the Victims of Our Own Success*, he writes that it's "not a coincidence that the end of the space age has coincided with a turning inward in the developed world, a crisis of confidence and an ebb of optimism and a loss of faith in institutions, a shift toward therapeutic philosophies and technologies of simulation, an abandonment of both ideological ambition and religious hope."[10] We're mired in economic stagnation, political stalemates, and cultural exhaustion. We're like kids who, after eating all of the Halloween candy, sit in a stupor on the floor asking, "What now?"

Douthat ends his chapter "Comfortably Numb" as follows: "If you want to feel like Western society is convulsing, there's an app for that, a convincing simulation waiting. But in the real world, it's possible that Western society is really leaning back in an easy chair, hooked up to a drip of something soothing, playing and replaying an ideological greatest-hits tape from its wild and crazy youth, all riled up in its own imagination and yet, in reality, comfortably numb."[11]

While Douthat doesn't say it explicitly, it seems that the primary and underexplored reason for our stagnation and decadence is mimesis. We lack a transcendent reference point outside the system. Meanwhile, everyone is more or less imitating everyone else. Our culture is stuck because we're fighting over space in a pool, next to the ocean. Yet nobody dares to talk openly about it, this mimesis. It's the hidden force driving our cultural development, and yet it's taboo to speak about, like envy.

It's as if everyone denies that the force of gravity exists and yet wonders why people keep falling.[12] No one dares to call themselves mimetic, or to point out the mimesis driving their decisions or beliefs or the behavior of those in their group.

Alexis de Tocqueville, the chronicler of America, described what sounds like a mimetic crisis of sameness in his *Democracy in America*, written in 1835. He saw the danger of naively conceived independence. What would happen in a society that was increasingly liberal and individualistic, with a high degree of equality, but in which differences between people were noticeable? It would run the risk of having an even greater degree of enmity between people than a society with less equality. "When all conditions are unequal, there is no equality great enough to offend the eye," Tocqueville wrote, "whereas the smallest dissimilarity appears shocking in the midst of general uniformity; the sight of it becomes more intolerable as uniformity is more complete."[13]

While we fight for equality in the areas that *do* matter—for fundamental human and civil rights, or for the freedom for each person to pursue

their thick desires (in the United States, this is called the "the pursuit of happiness")—we also begin fighting for equality in areas that do *not* matter, our thin desires: to make as much money as someone else, to have the same number of Instagram followers, to have the same amount of status or respect or professional prestige as any one of the nearly eight billion models on the planet.

The fight for things that do matter intersects and interacts with the fight for things that don't matter because mimetic desire has the effect of blurring the lines. It turns our attention away from thick desires and onto thin ones. When the desire for equality is hijacked by mimetic desire, the only things we see are imaginary or superficial differences.[14]

We find ourselves in a cycle of destructive desire. But that, in itself, is not deadly. It's deadly because people don't seem to think there's any alternative. Our society is decadent and stagnant because it lacks hope. Hope is the desire for something that is (1) in the future, (2) good, (3) difficult to achieve, and (4) possible. The fourth point is critical. Without the conviction that the fulfillment of a desire is possible, there is no hope—and therefore no desire. Hope is the soil in which thick desires grow. For lack of vision the people perish.[15]

In order to break out of this mimetic cycle, we're going to need to find something worth hoping for.

Instruments Versus Relationships

There are two approaches that people commonly take to escape from Cycle 1.

The first approach, *engineering* desire, is the approach of Silicon Valley, authoritarian governments, and the Cult of Experts. The first two use intelligence and data to centrally plan a system in which people want things that other people want them to want—things that benefit a certain group of people. This approach poses a serious threat to human agency. It also lacks respect for the capability of people to freely desire what is best for themselves and the people they love. The Cult of Experts, with their "Follow These Five Steps" approach to happiness, lacks respect for human complexity.

The alternative is the *transformation* of desire. The engineering approach is like extractive industrial farming, which uses pesticides and tills the land with large machinery, then measures success by seasonal yield, shelf life, and uniformity. The transformation approach is like regenerative farming, which can transform a barren piece of land into rich soil according to the laws and dynamics of the ecosystem. In our case, the ecosystem is one of human ecology—and desire is its lifeblood.

Transforming desire happens through relationships. Engineering desire happens in labs, with cold, lifeless instruments.

Engineering Desire

Technology companies have the power to engineer desire because they increasingly stand as mediators between people and the things they want. That is the definition of a mimetic model. Amazon mediates desire for things. Google mediates desire for information itself. Google started out simply as a search company, helping people find and access pages on the web. Within a few years, the company realized that its search results were not just data points about what people happened to be trying to find at any given time but early indications about what people wanted—information about their desires, which Google had access to before anyone else. Google pioneered what Harvard professor Shoshana Zuboff calls *surveillance capitalism*.[16] Companies that operate according to this model translate private human experience into behavioral data that can then be used to engineer their desires, or at least to exploit them for profit.[17]

On an earnings call in 2011, Google co-founder Larry Page explained Google's new mission as the transition from *search* to *satisfy*. "Our ultimate ambition is to transform the overall Google experience, making it beautifully simple, almost automagical, because we understand what you want and can deliver it instantly."[18]

Shoshana Zuboff recounts the following story in her book *The Age of Surveillance Capitalism: The Fight for a Human Future at the New Frontier of Power*. One morning in 2002, a team of Google engineers arrived at their desks to find that an odd phrase had surged to the top of the worldwide search queries: "Carol Brady's maiden name." Why the sudden interest in the family background of a character on a 1970s sitcom? The search query spiked at forty-eight minutes past the top of the hour for five consecutive hours.

It didn't take long for Google's engineers to figure it out. The game show *Who Wants to Be a Millionaire?*, watched by millions, posed the question to one of its contestants the night before. As the show rolled through different time zones, the question repeated at forty-eight minutes past the hour.

Because Google has access to leading-indicator data about people's desires, they're almost trading on insider information. In his 2018 book *Life After Google: The Fall of Big Data and the Rise of the Blockchain Economy*, technology guru George Gilder writes, "Google's path to riches is that with enough data and enough processors it can know better than we do what will satisfy our longings."[19] And that is true—as long as our desires are pedestrian and predictable.

Carol Brady's maiden name, in case you're wondering, is Martin. A two-second Google search could have told you that for free. Except it's not free. Every time we type something in the search box, we tell Google what we want. Sometimes we disclose things we'd never tell a soul. Google responds by giving us about 2,830,000 results in 0.59 seconds (at least that's what I got when I searched "limoncello chicken" while dinner planning just now). And in those 0.59 seconds, we've given Google our desires.

That's a very high price to pay.

CENTRALLY PLANNED WANTING

Political debates about the extent to which we should engineer desire are perennial. We just never frame them that way.

This is not the place for an in-depth exploration of desire and politics, but I will suggest one underappreciated way of looking at political questions: What effect does a political system or a policy have on *what people want*? What are its effects on desire?

Authoritarian regimes can only stay in existence so long as they can control what people want. We normally think of these regimes as controlling what people can and cannot do through laws, regulations, policing, and penalties. But their real victory comes not when they have authority over people's actions; rather, their victory comes when they have authority over their desires. They don't want to keep prisoners in cells; they want those prisoners to learn to love their cells. When there is no desire for change, their authority is complete.

The purpose of a "reeducation" camp is not about relearning how to write or read or interpret history, or even how to think; it's fundamentally about the reeducation of desire. Russian scholars Catriona Kelly and Vadim Volkov have pointed out in their essay "Directed Desires: Kul'turnost' and Consumption" that the transition to Soviet Russia came about through what they call *directed desires*. There was a subtle campaign to direct people to want certain things and reject others. The idea of *kulturnost*, or *culturedness* in English, began to emerge. It was a right way to live based on shared Russian cultural values.[20]

The leaders of the Gulag labor camps tried to starve prisoners' desires for things that were not aligned with their ideas of what was desirable. They fed those that were. Author Roy Medvedev recounts one woman's experience after being released: "I am disappointed in everything and believe in nothing anymore, but I have one desire—to eat ice cream every day, not beauty, not love."[21] Her desire for anything more than ice cream had been destroyed. But now she needed a refrigerator and economic security to eat ice cream all the time. She ended up supporting the very political party that had enslaved her because it was the only party that promised to help her fulfill the impoverished desire they had given her.[22]

Langdon Gilkey, a young American teacher at Yenching University during World War II, was arrested and interned at the Shantung Compound, a Japanese internment camp in present-day Weifang, China. He was interned there for two and a half years with businessmen, missionaries, teachers, lawyers, doctors, children, prostitutes—a cross-section of humanity. Gilkey was struck at the way the internment affected his desires: "I marveled at the way by which we can fool ourselves," he wrote in his 1966 book *Shantung Compound: The Story of Men and Women Under Pressure*. "We don some professional or moral costume so as to hide from ourselves our real desires and wants. Then we present to the world a façade of objectivity and rectitude instead of the self-concern we really feel."[23] The Shantung Compound had a disorienting effect on everyone's wanting—which made them that much easier to be secretly steered by those in charge.

Ideologies are closed systems of desire. They provide clear constraints about what is acceptable or not acceptable to want—whether it is the platform of a political party, the guiding ideology of a company, or the ideology that shapes a family system.

The salient feature of any ideology is the violence that it both covers up and constrains. In other words, an ideology keeps a group "safe" from

Tactic 13

LOOK FOR THE COEXISTENCE OF OPPOSITES

To cut through ideology, it helps to pay attention to what in Latin are called *coincidentia oppositorum*—the coincidence, or coexistence, of opposites: paradoxical figures, walking contradictions. People who are both meek and bold at the same time, humble and confident, or who utterly confound expectations. People or things or experiences that make us scratch our head and say, "Wait, those things are not supposed to coexist."

These coincidences of opposites point to something transcendent. The reason things *seem* like they shouldn't coexist is that they don't map onto how I experience the world. They don't have a place on my map of meaning, my mental model of how the world works. They are a sign that I need to go further, to reevaluate, to press deeper. They point to something beyond where I'm currently at.

Wise people have said that it's best to compare yourself only to who you were yesterday, not to who other people are today. That's a good start for escaping the trap of comparison and measurement.

But it's not enough. The person I was yesterday is not a model I can imitate. I can only look back at him (and usually, in my case, shake my head).

What I need is a model of the future—a transcendent model, where there are no more paradoxes of opposites, where there are no more contradictions, when my desires aren't in constant, unresolvable tension. We all do.

The coexistence of opposites is often a sign, pointing us in the right direction.

intruders who might bring with them an infectious strain of thought. There is no room for opposition. Girard once defined ideology as "the idea that everything is either good or bad."[24]

It's a sign of maturity to be able to hold on to two conflicting desires or two opposing ideas at the same time without immediately rejecting one or the other, before there has been time for a careful discernment. To live with desire is to live with tension.

AN EASY WAY OUT

In 1930, Polish writer Stanisław Ignacy Witkiewicz published a satirical novel called *Insatiability*.[25] In it, Poland is conquered by an army from Asia. The people are devastated, ruined—until they begin to hear about Murti-

THE COMPARISON GAME

Bing, a philosopher from the conquering army who has found a way to deliver humans a new philosophy of life in a single pill.

Soldiers soon show up on street corners slinging the "Pill of Murti-Bing." The Poles become hooked on the new drug. The pills are a way of engineering their desires so they can easily accept their new way of life.

But because the pills are not an organic development of their thoughts and desires, the people who take them develop split personalities. They go crazy. They're divided against themselves.[26]

The Pill of Murti-Bing is a forerunner of the *Matrix* movies and has some similarities to the drug soma in Aldous Huxley's *Brave New World*. People's desires are artificially shaped by external forces. We should face the real possibility that we will soon have, or already do have, Pills of Murti-Bing.

Will you take one?

Transforming Desire

There are two different ways of thinking that correspond, respectively, to engineering desire and transforming it: *calculating thought* and *meditative thought*. I draw these distinctions loosely from the work of philosopher Martin Heidegger.[27]

Calculating thought is constantly searching, seeking, plotting how to reach an objective: to get from Point A to Point B, to beat the stock market, to get good grades, to win an argument. According to psychiatrist Iain McGilchrist, it's the dominant form of thought in our technological culture. It leads to the relentless pursuit of objectives—usually without having analyzed whether the objectives are worthy to begin with.[28]

A monk in charge of training novices at a monastery told me that in recent years he has noticed that young postulants (men on their way to becoming monks) bring stacks of books with them when they pray in the chapel. They are habituated to think that without "input" there can be no "output." The hypertrophy of calculating thought is a product of our technological development—humans imitating machines.

Calculation also reflects an engineering mindset. Calculating leaders deal with desires by building an algorithm to better predict them, creating an app to nudge them in one direction or another, or constructing an inorganic "company culture" to form them. I sometimes wonder how top-down company culture cults are any different than the phenomenon of *cuius regio, eius religio* (whose realm, whose religion) of the Holy Roman Empire, in which different princes or rulers had the right to enforce their preferred religious beliefs on the population.[29]

There's nothing wrong with engineering. But some things are meant to be engineered (motorcycles) and other things are not (human nature).

Calculating thought has become the primary mode of thought, often to the exclusion of meditative thinking altogether, which leads eventually to forms of social engineering, technological manipulation, and the loss of empathy. "From the massacre of the Armenians, to the horrors of the Shoah and Cambodia, and the crimes in Rwanda, whole peoples have been coldly murdered, sometimes even with bureaucratic zeal," writes Girard.[30] The calculating mindset allows the scapegoat mechanism to thrive.

Tactic 14

PRACTICE MEDITATIVE THOUGHT

August Turak is the author of the award-winning book *Brother John: A Monk, a Pilgrim and the Purpose of Life*. In the early 1980s, he was a sales executive at MTV when the network was just getting off the ground. When I visited him at his ranch in North Carolina, he told me the following story.

Turak was riding the New York City subway with a fellow MTV executive, a brilliant thinker who liked to give his friends and colleagues puzzles. He challenged Turak: "Give me the next number in the series; 14, 18, 23, 28, 34."

He prided himself on being good at solving puzzles. "I'm racking my brain," he tells me. "Eighteen minus fourteen equals four; twenty-three minus eighteen equals five, and so forth. But I can't figure it out." He finally gave up.

With that, his colleague pointed to the big "42" on the wall of the 42nd Street subway station, where their train had just stopped. The numbers referred to the stops the train had made up to that point: 14th Street, 18th Street, 23rd Street, and so on. "I was shouting out all these calculations," he said. "All the while, I'm looking at these signs with the numbers he gave me in the puzzle, and I still didn't see it."

He was calculating, and it caused him to miss what was right in front of his face. Meditative thought helps us sink down into reality and notice *divergent* possibilities rather than *converging* on one ("It's a math problem"). Meditative thought is also essential to the process of discerning desires.[31]

The best way to get started practicing meditative thought is to pour yourself a beverage and look at a tree for an hour. An entire hour. There is no goal in this exercise other than learning how not to have a goal. As you look at the tree, pay attention to everything you notice. You should find that your calculating thought slowly gives way to meditative thought. If not, repeat as necessary.

Meditative thought, on the other hand, is patient thought. It is not the same thing as meditation. Meditative thought is simply slow, nonproductive thought. It's not reactionary. It's the kind of thought that, upon hearing news or experiencing something surprising, doesn't immediately look for solutions. Instead, it asks a series of questions that help the asker sink down further into the reality: What is this new situation? What is behind it? Meditative thought is patient enough to allow the truth to reveal itself.

Meditative thought opens the door to transformation. When the calculating, processing part of our brain calms down, the meditative part—which takes in new experiences—is given the ability to work, integrating those new experiences into a new framework for reality.

The calculating brain is only able to fit new experiences into *existing* mental models. The meditative brain *develops new models*. If we spend all of our time in calculating mode, we spend our lives trying to fit every new encounter into boxes. And when it comes to desire, that's deadly.

Both modes of thinking are useful in different circumstances. If I'm making moves in the stock market, I should exercise calculating thought. If I'm trying to make sense of a new and unexpected situation in the world, or the discovery of thick desires, I need meditative thought. Calculating thought simply doesn't stay in the present long enough for anything thick to present itself.

Meditative thinking is the antidote to a culture of hyperspeed mimesis because it allows time to develop thick desires. Transformation happens when I spend enough time with my desires to know them by name and know whether or not I want to live with them.

Calculating thought is the more mimetic form of thought. But mindset alone is not enough. We're in relationships with others. And relationships are where mimetic desire lives.

Pivotal Spaces

Many relationships are held together by mimetic bonds: between players who compete for a coach's respect, colleagues who compete for status, and academics building out their CVs.

Mimetic tension is present even in relationships that are, on the whole, healthy: between spouses, parents and children, or colleagues. Even your relationship with your best friend might be—and probably is—tinged by mimesis. Healthy competition can be good; here, we're talking about mimetic rivalry. The key is recognizing the ways in which a relationship is marked by mimetic rivalry, and confronting them.

Transforming desire involves changing the nature of our relationships. Let's start with three places where most of us spend the majority of our time: our family, our imagination, and our work.

THE FAMILY

Families are the place where people first learn how and what to want.

When we are children, our entire menu of desires—the things that we can choose to want or not want—is mostly limited to the objects and roles that our family presents to us and then rewards us for playing. Those roles may include being an obedient child for an emotionally needy parent, living up to the model set by older siblings, being a good liberal or conservative, being religious or atheistic, or any number of other things that the family value system is set up to shape.

When children are very young, parents are their only mimetic models. What the parents want, the child wants. Older siblings come next. But it doesn't take long—typically until the age of three, or the point at which they realize their parents are not gods, whichever comes first—before they start looking around for other models. And just about any model will do.

According to a 2015 article in the *Wall Street Journal* by Jacob Gershman, two-year-old Grayson Dobra of New Orleans became fascinated with the commercials for personal injury lawyer Morris Bart. As soon as Grayson was able to talk, he started blurting out, "Bart! Bart!" So, on his second birthday, his mother threw him a Morris Bart–themed party: a Morris Bart–themed cake, a Morris Bart cutout figure, and Morris Bart–inspired gifts. Grayson found his first model outside of the family—in Celebristan.

By the time children are teenagers, they've left their childhood models behind. Adolescence inaugurates a hyper-mimetic phase that leaves even the most grounded kids with whiplash. Each is trying to answer fundamental questions: Who am I? Who do I want to be?

Through it all, parents can help their children recognize which of their desires are thick and which of them are thin, and encourage them to cultivate the thick ones. They can do this by emphasizing things that might lead to fulfillment (for instance, pointing out that their child's amazing piano concert from last year seems to have taken their love for music to another level) and deemphasizing those that don't (such as putting the anxiety they feel about getting an A- because their best friend got an A into the proper perspective).

Most importantly, parents have a responsibility to model healthy relationships. That means paying careful attention to their own mimetic impulses. Even in ways that seem innocent or insignificant. Reacting mimetically to every piece of political news at the dinner table, or to every minor injustice that a child suffers in school or sports, or using kids as pawns in a rivalry with other parents (such as buying your kid a nicer car than their friend's parents could afford to buy her, to signal your status)—all these things create an atmosphere in which mimetic behavior is learned and becomes the norm.

Most people tend to be about as mimetic as the people around them. The mimetic behavior of parents tends to be learned and adopted by their children. Often, so do their scapegoats. We should be mindful of who our children learn to love and hate.

THE IMAGINATION

What do blind people dream about? The answer depends on when they lost their sight. A person who went blind at age eight can dream using all the sensory inputs their brain received while they could see. People blind from birth are different. They don't dream in images because their brain has no images to work with. Instead, they dream in sensations and sounds (falling into a manhole and getting hit by an unseen car are common). In short, we can only dream using the inputs we've received.[32]

When it comes to our desires, we usually behave as if we were blind. We look to other people who we think can "see" better than we can—our models—in order to learn what is worth looking at and pursuing. Each of us has a universe of desire that is only as big as our imagination.

How have our imaginations been formed?

Much of life is made up of tacit knowledge—what the philosopher Michael Polanyi called "inarticulate rationality." These are things we know but can't explain. We know many things that we'd stumble trying to communicate clearly to another person—even ourselves.[33] I experienced this when I tried to teach my wife how to snowboard for the first time. It wasn't pretty. And I'm not referring to her snowboarding, but to my bumbling attempt to teach.

We strapped in at the top of the bunny hill. I hopped up and instinctively shifted my weight to my heel edge to keep from sliding down the small gradient we were on (my first mistake: not finding level ground for the lesson). "Here, just lean . . ." *Boom.* Before the words had come out of my mouth, Claire tried to get up and had already landed on her butt. We suffered through the next hour with me trying to give descriptions of what I was doing, and none of it working. Then she figured out how to shift her weight on her own, through trial and error. She told me all of the simple cues I *could've* given her to save her from fifty falls in sixty minutes. But the truth is that I had no idea how I did what I did. I didn't remember what it was like when I first learned.

I recalled the fable of the centipede. One day a spider watched a centipede, fascinated by his dexterity. The spider asked him how he manages to coordinate the movement of all one hundred of his legs at the same time. The spider only has eight legs; he couldn't imagine having to move an extra ninety-two. "Um, well . . . let's see," said the centipede. "I move this one first . . . no, wait, this one . . . no, maybe this one . . . and then I . . . no, that's not right." The centipede balled up just thinking about it. He had tacit knowledge.

Fluency in a new language, a sense of humor, emotional intelligence, and aesthetic sensibility are all things we probably know tacitly but cannot articulate fully. So is a vibrant imagination, which is filled with models of desire at a young age.

From the time children hear their first fairy tale, their imagination takes off with fantastic images of noble ideals and adventure: heroism, sacrifice, beauty, love. These are all things central to the fiber of our humanity, and yet we have a hard time explaining why they're important.

Literature is one of the primary influences on the imagination—a school of desire. Literature is where young minds enter into the stories of others' desires, real or fictional. Sure, it exposes children to mimetic forces and often inflames their own desires (your kid who reads the Harry Potter books wouldn't mind being a wizard for a day)—but think of it as a training ground for dealing with and discerning which desires lead to where. In good novels, it plays out in the stories.

Our desires are only as big or as small as the models that we are exposed to. Fictional characters who model great, thick desires can be a counterbalance to real-life models of weak, thin desires.

Education has shifted away from the liberal arts and toward increasingly specialized, technical knowledge—calculating thought. How will this affect the formation of desire in future generations? We don't know. But we should think seriously about how our systems of education are shaping students' imaginations, and therefore desires.

WORK

I believe the purpose of work is not merely to make more but to become more. The value of work cannot be measured by the objective output of a job alone; it must take into account the subjective transformation of the person who is working.

Two doctors could make exactly the same rounds in the same hospital. After ten years, one could grow bitter and resentful as a result of the long hours, bad cafeteria food, broken insurance system, and ungrateful patients; the other could experience the same things but use them to become a more caring, patient, and understanding human being.

Employers have a responsibility to think about this subjective dimension of work. How does the company and the nature of the work within it contribute to the overall flourishing of the person?

In 2015, Dan Price, the founder and CEO of Gravity Payments, voluntarily forfeited most of his nearly $1 million annual salary so that he could boost the minimum salary at his company to $70,000 over the next three years. At the time of his decision, the average salary at the company was $48,000. The lower salaries were at the level the market supported, the salaries that his competitors were paying for similar jobs. But in pricey Seattle, Washington, they didn't support much more than just getting by. Many employees didn't feel stable enough to start families.

In the five years after Price made the decision, the company prospered. The company went from processing $3.8 billion to $10.2 billion in transactions. More importantly, employees were thriving. The headcount more than doubled, and employees were able to pursue their thick desires—like having children. Before the minimum salary adjustment, there were between zero and two babies born to Gravity employees each year. After the adjustment, the number was around forty.

Sometimes the market isn't a good indication of what people want. It's good at price discovery for thin desires, but not necessarily for thick ones.

Transforming desires at work doesn't happen by tinkering with the status quo. It happens when someone steps outside a mimetic system—for instance, the "industry standard" in compensation—and takes a more comprehensive view of life and humanity.

There are many new and different ways to reward good work that fall outside the stale Marxism-versus-capitalism frame, but few entrepreneurs are willing to explore them.

Businesses also have the power to make what is more sustainable more desirable. Sadly, most companies don't think of their mission in terms of how they shape desire. For every company profiting from unsustainable practices, what if there were two more creating sustainable opportunities and making them attractive?

Yolélé Foods imports and introduces the ingredients and cuisines of West Africa to the United States.[34] Its flagship ingredient is fonio, an ancient, drought-resistant grain that has been grown in the Sahel region for millennia. But many people in Senegal and other parts of West Africa look down on it. Outside of its growing regions, fonio isn't recognized as having much value anymore—and that perception of value is largely mimetic.

Yolélé's co-founder, Senegalese chef Pierre Thiam, told me that there is a common perception in Senegal that "what comes from the West is best." Anything produced locally is perceived to have diminished value. When I sat down with him in 2019, Thiam explained to me how this idea was instilled in the minds of the Senegalese by their colonizers. Because the Senegalese were forced to grow monocrops such as peanuts for export, local crops were displaced, and France had to import broken rice from what was then Indochina to feed the Senegalese, who no longer grew their own food.

Yolélé Foods is intentional about making the image of West African ingredients and foods popular in the United States so that people in West Africa will have a renewed desire—and the financial incentive—to grow and eat fonio and other local products.

Every business should think seriously about how its mission is aligned with models of desire.

The Three Inventions

Toward the end of his life, René Girard grew increasingly worried that we faced a future in which there would be more mimesis—wars with no definitive end, technology fueling our most mimetic instincts, and globalization as an accelerant of mimetic crises—with no effective means of controlling the conflict.

There have been two major social inventions in history that mitigated the negative consequences of mimetic desire: the scapegoat mechanism and the market economy. Could there be a third?

THE EVOLUTION OF DESIRE

THE FIRST INVENTION

The scapegoat mechanism prevented a society in crisis from destroying itself from within. It worked in a paradoxical way: the scapegoat mechanism contained violence *through* violence. Instead of a war of all against all, there was a war of all against one. In spite of its injustices, Girard recognized that the scapegoat mechanism had a stabilizing effect in early societies.

In modern Western civilization, the scapegoat mechanism has grown less effective, like a drug to which we've become desensitized. Its weakness is apparent in the twenty-four-hour news cycle, reality television, and social media. It only takes days and in some cases hours after a scapegoat's destruction before more blood or ridicule is called for.

We saw in Chapter 4 that this lack of effectiveness is a result of the scapegoat mechanism having been exposed. We no longer fully believe that what we are doing is just. The loss of our belief in the scapegoat's guilt led to its impotency. Scapegoats are like the gods in Neil Gaiman's *American Gods*—they only exist so long as people believe in them.

THE SECOND INVENTION

As the scapegoat mechanism has lost its effectiveness, the modern market economy has arisen to take its place.[35]

The market economy channels mimetic desire into productive activities.

When the talented CTO of a billion-dollar start-up becomes jealous of the co-founders, he doesn't lead a coup d'état to put their heads on stakes—he leaves and starts his own company. Nikola Tesla and Thomas Edison didn't compete for land or physical dominance; they competed for prestige. And those options, on the whole, are a good thing compared to the alternative, which is direct and violent physical conflict.

Economic competition is less bloody than the sacrificial world that it supplanted. At the same time, it produces its own victims: the poor who don't have access to markets, exploited workers, and winner-take-all systems. It exacerbates the differences between those who are in and those who are out.

Despite consumerism's many problems, it channels rivalry and desires into places where the negative effects accrue primarily to the people who indulge their thin desires. If you only eat at Del Frisco's steakhouse, you might put on weight and become a boring conversationalist, but at least you aren't engaging in swordfights in the street to preserve your social status.

As long as we are competing to have the nicest car and house in the neighborhood, we're not trying to annex our neighbor.

THE THIRD INVENTION

The scapegoat mechanism was the first major social invention to address the problem of desire. The market economy was the second. Neither will be able to protect us from mimetic escalation and crises in the future.

The preservation of humanity might rely on a third invention—one that is yet to be discovered, or that is in the process of being discovered. Humans will have to find a new way to channel desire in productive and nonviolent ways. Without one, mimetic desire will spiral out of control. It's impossible to know what this emergent social mechanism might be, but I'll indulge in some brief speculation.

Humans could create a technological superstructure that plays the same role that archaic religions played in scapegoating: diffusing violence—through billions of bits and bytes—into the ether. It might be an invention that brings about an evolution in money, which makes it easier to measure and reward value creation at the individual level. It might be an invention that accelerates space exploration and colonization so that humans have entirely new parts of the cosmos to explore and therefore will be less focused on destroying each other. Or it might be an invention in education that makes it easier for each person to carve out their own track.

Is it possible that the innovation has already happened with the advent of the internet? When something goes wrong, people instinctively search for answers on Google. The internet, standing in for crowd violence, renders verdicts and directs people to a thousand different places—outlets where they can find a form of catharsis uniquely tailored to them, whether that be in a Reddit forum or a Facebook post.

I'm doubtful that the internet is the third invention. It seems to me to be exacerbating more violence than it quells.

In the absence of a new social invention, all we can do is what we can do. That starts with how we shape and manage our own desires.

Single Greatest Desire

Naval Ravikant, the founder of AngelList, said by the *Wall Street Journal* to have a personal philosophy of "rational Buddhism," is one source of thoughtful reflection for many entrepreneurs. Few tech CEOs have spoken more directly about the role of desire in business and in life.

"Desire is a contract that you make with yourself to be unhappy until you get what you want," he said.[36] Ravikant is drawing on the perennial understanding of numerous spiritual traditions about the link between desire and suffering: desire is always for something we feel we lack, and it causes us to suffer.

Mimetic desire manifests itself as the constant yearning to be someone or something else (what we called *metaphysical desire*). People select models because they think the models hold the key to a door that just might lead to the thing they have been looking for. But as we've seen, this metaphysical desire is a never-ending game. We cycle through models faster than we cycle through clothes. The act of winning, of gaining possession of the thing that the model made us want, convinces us that we chose the wrong model in the first place. And so we go in search of another one.

Mimetic desire is a paradoxical game. Winning is how you lose. Every victory is Pyrrhic.

The world is full of models. The business world is particularly fond of them. There are financial models. There are best practices, guidebooks, templates, and daily blogs claiming to offer road maps to success. There are role models on the covers of magazines and in the pages of the *Wall Street Journal*. There are books. Everyone is either a model or signaling that they should be.

They all offer external frameworks, or schemas, of desire. This seems to be what people are looking for. When my students ask me what they should do with their lives, they are always looking for me to hand them a playbook. They want a road map. I tell them about Dave Romero showing up on my doorstep in Las Vegas. I ask them if the Stern School of Business should give me my money back for not teaching me how to deal with him.

All schemas ultimately come up short. Life is about navigating an uncertain future, and every one of our current schemas is inadequate.

The English word "schema" comes from Greek. It forms the root of the Modern Greek verb *suschématizó*, which means "to conform to." For instance, the Greek phrase "*Me syschematizesthe!*" means "Don't conform!" More specifically, it means something like "Don't fit yourself according to the pattern of any external model."

An engineered desire, by definition, is always according to a model. No architect begins a blueprint without a schema in hand.

The transformation of desire, on the other hand, is a dynamic process. The Greeks have an entirely different word for a total transformation from within, one that isn't wholly patterned on any one particular model: *metamorphosis*.

This kind of substantial transformation—which is, at root, a transformation of desire—is painful. One thing that every spiritual tradition is clear about is that changing how we desire, at least in a positive way, requires suffering. Nobody wants to let go of thin desires.

"Pick your one overwhelming desire. It's okay to suffer over that one," Naval Ravikant said on Joe Rogan's podcast. Other desires must be let go of.[37]

Letting go of lesser desires is a kind of death.

When I think about plugging into a single greatest desire, I think of one of my favorite authors, the American writer Annie Dillard. In *Teaching a Stone to Talk: Expeditions and Encounters*, a collection of her essays, she writes about waking up from a self-absorbed childhood and becoming immersed in the present moment of the larger world around her. Later in life, she wrote a poetic piece of nonfiction called "Living Like Weasels." It's about a chance encounter with a weasel—yes, a weasel—in nature. I come back to it frequently because of the beauty of Dillard's words and her attentiveness to how much a weasel has to teach us about life. I can't say it better than Dillard, so I am going to quote her a bit.

Annie Dillard tells the story of a man who shot an eagle out of the sky. When the man examines the bird's body, he notices the bones of a weasel jaw locked firmly around the eagle's neck. It must have swooped down to snatch the weasel off the ground. But the weasel, with perfect timing, turned its head around at the last moment and sank its teeth into the eagle's neck.

The weasel tightened its grip on the flesh of the eagle's neck as the bird flew high into the sky until eventually—who knows how long that weasel was dangling from the eagle's neck?—either the eagle or the wind picked the weasel's dry bones apart and there was nothing left but the remnant of a jaw.

"The thing is to stalk your calling in a certain skilled and supple way, to locate the most tender and live spot and plug into that pulse," wrote Dillard. "I think it would be well, and proper, and obedient, and pure, to grasp your one necessity and not let it go, to dangle from it limp wherever it takes you. Then even death, where you're going no matter how you live, cannot you part."[38]

We're not guided entirely by instincts like the one that helped the weasel plug into that pulse. But we must make a decision about what it is that is worth sinking our teeth into. Otherwise, our bones will get picked dry by the winds of mimetic forces without our ever having staked a claim on anything that touches us at the depths of our being.

Stalk your greatest desire. When you find it, let all of your lesser desires be transformed so that they serve the greatest one. "Seize it and let it seize you up aloft even," writes Dillard, "till your eyes burn out and drop; let your musky flesh fall off in shreds, and let your very bones unhinge and scatter, loosened over fields, over fields and woods, lightly, thoughtless, from any height at all, from as high as eagles."[39]

Taking hold of your greatest desire necessarily means taking hold of models. We can't access our desires without models. And we will always follow those models that are most real to us—who possess a quality of life that we feel transcends our own.

So stalk your highest and noblest desire, but you will have to find it in the form of a model. On this particular day, as you read these words, it might be a character from a book, a leader, an athlete, a saint, a sinner, a Medal of Honor winner, a love, a marriage, a heroic action, the greatest ideal you can possibly conceive.

But that model will not be the end. Because it's exterior, it can't automatically effect the inner transformation that needs to happen in order for you to transcend the model. If the inner transformation doesn't happen, we get stuck playing a never-ending game of whack-a-mole when it comes to our models and desires. When the inner transformation happens, there's a reflexive process that helps the thin desires fade away and the thick ones take root.

There is no opposition between external models, or schemas, and the inner transformation, or metamorphosis. The point is to make sure that as you pursue a model, the pursuit is simultaneously effecting the inner transformation that will help you select new and better models.

LOVE AND RESPONSIBILITY

Thin mimetic desires abound. They're peddled to us every minute of the day. We can nip at them, maybe even sink our teeth into them, but they won't take us where we want to go.

Our choice is between living an unintentionally mimetic life or doing the hard work of cultivating thick desires. The latter may require us to suffer from the fear of missing out on the shiny mimetic objects that surround us.

At the end of my life, I believe the primary thing I'll fear having missed out on is the pursuit of thick desires. Desires that I'll feel satisfied about having poured myself out for. If I'm going to die of exhaustion—and, eventually, all of us will—it's not going to be from chasing thin desires. It's going to be from grabbing hold of a thick one and holding on until there's nothing left.

The destructive mimetic cycle works when people are convinced of the absolute primacy of their own desires. They're even willing to sacrifice others in order to fulfill them. But in the positive cycle of desire, people respect the desires of others as they would their own. What's more, they take an active role in collaborating with others to help them achieve their single greatest desire. In a positive cycle, we are all in some sense midwives to the thick desires of our neighbors.

The simplest definition of love is wanting what's good for another.

Tactic 15

LIVE AS IF YOU HAVE A RESPONSIBILITY
FOR WHAT OTHER PEOPLE WANT

Through our relationships, we help other people with their wants in one of three ways: we help them want more, we help them want less, or we help them want differently.

There is no person we encounter—not even in the most uninteresting interaction of our day—whom we do not help desire in one of these three ways. The changes are usually imperceptible. But like a giant flywheel, we are gently nudging other people's desires in one or another direction.

Living with an awareness of mimetic desire brings with it the responsibility to defuse rivalry and to model positive desires in small ways daily.

Italians have a way of saying "I love you" that is particularly instructive: *Ti voglio bene*, they say. It means "I want your good"—I want what's best for you.

We have a responsibility to shape our own desires. As we've seen, we can't do that without others. The duty to shape our desires goes hand in hand with the responsibility to care for the relationships that we have with others.

The transformation of desire happens when we become less concerned about the fulfillment of our own desires and more concerned about the fulfillment of others. We find, paradoxically, that it is the very pathway to fulfilling our own.

The positive cycle of desire works because the primary thing being imitated is the gift of self. This is the positive force of mimetic desire behind every beautiful marriage, friendship, and act of charity.

In the end, wanting is another word for loving. And that, too, is mimetic.

"We can live any way we want," writes Dillard near the end of her essay. "People take vows of poverty, chastity, and obedience—even of silence—by choice." She noticed a clear difference between herself and the weasel: "The weasel lives in necessity and we live in choice, hating necessity and dying at the last ignobly in its talons."[40]

Our choice is to yield to the mimetic forces making claims on our desire at every moment or to yield to the freedom of our single greatest desire: doing the one thing that we were made to do, all of the time, over and over and over again, until we've developed a desire thick enough to stake our life on.

In the meantime, and probably at all times, we have something warm to sink our teeth into: wanting what we already have.

AFTERWORD

René Girard wrote that "the author's first draft is an attempt at self-justification." First drafts of anything—whether books or companies or relationships or life plans—are often about coming to grips with what we want.[1]

Girard believed that the best novelists read their first drafts and see right through them. They see that the first draft was a "put-up job"—an unconscious attempt to deceive their readers and themselves about the complexity of their desires. (Stephen King has written that the most important thing he learned from Carrie White, the lead character in his first horror story, is that "the writer's original perception of a character or characters may be as erroneous as the reader's.")[2]

The experience of reading that first draft devastates and disillusions the author, striking a blow at their pride and vanity. "And this existential downfall is the event that makes a great work of art possible," Girard said.[3] The writer begins again—but this time without the Romantic Lies that blinded them to their own mimesis.

Before this, the writer's characters were either good or evil. After this, there's nuance. The characters have to grapple with mimetic desire and rivalry. The author sees that life is a process of ever-evolving desire.

If I mentioned your name anywhere in this book, whether in a positive way or in a critical way, you are probably some kind of model for me. As you've affected my desire to write a book about mimetic desire, I hope that I've affected someone else's desire to write a better one.

I'm probably already hard at work, competing with you.

ACKNOWLEDGMENTS

This book was built on the broad and lofty shoulders of René Girard, but many others lent me theirs, too—as well as their eyes and ears, and in some cases their desires. I couldn't have gotten to the finish line without them.

My wife, Claire, heard the word "mimetic" more times in one year than any person should have to hear in five lifetimes. She, more than anyone else, was the sounding board for my wildest ideas. Some of those, thank God, did not make it into this book. That is thanks to her. She was also a tireless and smart editor and interlocutor who encouraged me and contributed more than anyone to bringing this project to fruition.

Some of the ideas in this book are derivative of the work of others. I imitated them. So thanks to these models: Jim Collins, whose analogy of the flywheel clarified my own thinking about cycles of desire; Nassim Nicholas Taleb, whose language of "Extremistan" and "Mediocristan" was a model for "Celebristan" and "Freshmanistan"; the many Girard scholars and practitioners whose thinking over the past five decades shaped my own, especially Paul Dumouchel, Jean-Pierre Dupuy, James Alison, Cynthia Haven, Martha Reineke, Sandor Goodhart, Andrew McKenna, Suzanne Ross of the Raven Foundation, Steve McKenna, Ann Astell, Gil Bailie (who deserves credit for the term "disruptive empathy," which I merely elaborated on), and Wolfgang Palaver.

I'm grateful for Jim Levine's support throughout the entire process. He's the best literary agent that I could ever hope for, and I thank Adam Grant for introducing me to him. Jim has been a wise mentor and totally unshakable and consistent despite working through a global pandemic.

At St. Martin's Press, Tim Bartlett was like a great sports coach who knew just the right cues to give me at just the right times to get the best out of me. He saw the book's importance and ushered it through to the end with a deft touch. To everyone else at St. Martin's Press (too many to name individually) instrumental in bringing this project to completion: thank you. I'm proud to be your author.

Megan Hustad was invaluable for her insights and organization of the manuscript. My colleague Rebecca Teti displayed the same grace, wisdom, and fortitude that she does on a daily basis. Others who helped make this book what it is: Rod Penner, Brian Williamson, and the rest of the folks at Pruvio; Ben Kalin; my tireless and amazing Swiss Army knife of an assistant, Grady Connolly; Christine Sheehan; and all of the good people at bars and restaurants and cafés in Washington, DC, New York City, and around the world—many of which, sadly, are now closed—where I often sat during late nights and early mornings working on this manuscript.

A special thanks to Liana Finck, whose illustrations add life to these pages. Working through ideas with her and finding ways to portray them visually was one of the most rewarding parts of the entire process. Thinking through drawings affected my writing in a positive way. I was grateful for the opportunity to work with someone so thoughtful, talented, and kind.

Thank you to my colleagues, partners, and friends for helping to refine my thinking: Dr. Joshua Miller, Andreas Widmer, Frédéric Sautet, Tony Cannizzaro, Michael Hernandez, David Jack, Fr. Brendan Hurley (SJ), John Souder, Michael Matheson Miller, Carlos Rey, Gregory Thornbury, Anthony D'Ambrosio, Louis Kim, Brandon Vaidyanathan, and many others whom I can't properly thank in this way.

Thanks also to all those who contributed to the ideas in this book and who made themselves available for conversations: Chef Sébastien Bras, Peter Thiel, Jimmy Kaltreider, Trevor Cribben Merrill, Chef Pierre Thiam, Imad and Reem Younis, Dean Karnazes, Aimee Groth, Dr. Andrew Meltzoff (who was especially helpful), Mark Anspach, Bruce Jackson (who courteously supplied the photographs of Girard), Dr. Roland Griffiths, Naresh Ramchandani, Tyler Cowen, Dan Wang, Jonathan Haidt, and others whom I've either neglected or unjustly omitted due to space.

Last, thank you to my parents, Lee and Ida Burgis, and to my grandmother Verna Bartnick for giving me the gift of life, faith, hope, and love. AMDG.

But by now my desire and will were turned, like a balanced wheel rotated evenly, by the love that moves the sun and the other stars.

—Dante Alighieri

And there will be others, in any case, who will repeat what we are in the process of saying and who will advance matters beyond what we have been able to do. Yet books themselves will have no more than minor importance; the events within which such books emerge will be infinitely more eloquent than whatever we write and will establish truths we have difficulty describing and describe poorly, even in simple and banal instances.

—René Girard

APPENDIX A:
GLOSSARY

Terms invented or used in a specific sense in this book are indicated with an asterisk (*)

Anti-mimetic*
A person, action, or thing that counteracts the negative forces of mimetic desire. It is a specific way of being countercultural—less like a hipster, more like a saint.

Celebristan*
The world of external mediation.

Core motivational drive
A unique and enduring behavioral drive orienting a person to achieve a distinct pattern of results. Understanding core motivations can help people identify their thick desires and better align their desires with their motivational drive.

Cycle 1*
A process of destructive desire that leads to conflict.

Cycle 2*
A process of constructive, value-creating desire.

Desire
A complex and mysterious phenomenon of human life in which people are drawn toward certain people and objects that they believe are worthy of pursuit. Desires are different than needs because they require a model. "Man is the creature who does not know what to desire, and he turns to others in order to make up his mind," wrote René Girard. Desire is what leads humans to seek transcendent things.

Discernment
A process for making decisions that includes but goes beyond rational analysis. It comes from the Latin word meaning "to distinguish one thing from another." Discernment involves a power of perception, tacit knowledge, and the ability to read desires. Because desires lack scientific and objective criteria to judge them

by, discernment is required in order to decide which desires to feed and which to starve.

Disruptive empathy
Empathy that disrupts Cycle 1.

Double bind
When an imitator and a model end up taking *one another* as models, so that each is both an *imitator of* and *model to* the other.

External mediation
When a person imitates the desires of someone separated from them by time, space, or social sphere and there is little chance of the person ever coming into contact with the person they are imitating, their model. In external mediation, the model mediates desire from *outside*, or external to, the world of the subject.

Freshmanistan*
The world of internal mediation.

Fulfillment Stories
Stories about times in a person's life in which they took action they believe they did well and which brought them a deep sense of satisfaction. Fulfillment Stories help reveal a pattern of core motivational drive.

Hierarchy of values
A system of values understood as not necessarily equal, but interrelated and forming part of a unified whole.

Imitation
Taking someone or something else as a model for action. Children are experts at imitation; adults usually mask it. Imitation is the positive force driving childhood development, adult learning, and the acquisition of virtue. Imitation is neutral—we can imitate either positively or negatively.

Internal mediation
When a person lives within the same time, place, or social sphere as their model of desire and the likelihood of coming into contact with the model is high. In internal mediation, the model mediates desire from *inside*, or internal to, the subject's world.

Mediation of desire
The process through which desire takes shape in the dynamic relationship between a subject and a model.

Meme (Memetic) Theory
Not to be confused with the mimetic theory that originated with René Girard. The field of memetics studies how information and cultures develop based on prin-

ciples of Darwinian evolution. The term "meme" was coined by ethologist and evolutionary biologist Richard Dawkins in his 1976 book *The Selfish Gene*. It was meant to evoke the word *gene* because a meme is the cultural equivalent. Memes are words, accents, ideas, tunes, and more that spread from brain to brain through some process of replication or imitation.[1]

Mimesis
A sophisticated form of imitation that in adults is usually hidden. In mimetic theory, mimesis has a negative connotation because it usually leads to rivalry and conflict—that's one of the main reasons Girard differentiated it from common imitation by calling it "mimesis," from the Greek word for "imitation." People are more conscious of imitation than they are of mimesis. Mimesis can be positive or negative, but it is usually negative because it is denied or disguised.

Mimetic crisis
A mimetic crisis occurs when rivalrous mimetic desire has spread through a community, leading to undifferentiation. The result is chaos that threatens to tear the community apart socially.

Mimetic desire
Desire generated and formed through the imitation of what someone else has already desired or is perceived to desire. Mimetic desire means that we choose objects due to the influence of a third party, a model or mediator of desire.

Mimetic rivalry
Mimetic desire that has progressed to competitive rivalry—two parties who want the same thing compete for an object.

Mimetic systems
Systems that run on and are sustained by mimetic desire.

Mimetic theory
An explanation of social and cultural phenomena based on the role of imitation in human behavior—particularly the imitation of desire (mimetic desire) and its consequences. The theory explains the relationship between mimetic desire, rivalry, violence, the scapegoat mechanism, and the religious and cultural rituals, taboos, and prohibitions designed to prevent mimetic crises.

Mirrored imitation*
When a person tries to differentiate themselves from a mimetic rival by wanting something different or the opposite of whatever the rival wants.

Misrecognition
In mimetic theory, misrecognition (or "mis-knowing") refers to the tendency of people or groups caught up in the throes of mimetic desire to have their perception distorted and to misidentify people or things as the cause of their problems.

Misrecognition enables the effectiveness of the scapegoat mechanism. The concept of misrecognition—or *méconnaissance*, in the original French—is a critical one throughout Girard's work, but the word is hard to translate into English without losing some of its original meaning. The best treatment of the concept comes from philosopher Paul Dumouchel in his essay "De la méconnaissance" from his book *The Ambivalence of Scarcity and Other Essays*.

Model
A person, thing, or group that shapes and orients the desires of another.

Motivational pattern
The pattern of core motivational drive revealed in Fulfillment Stories. A person's motivational pattern is the thread that runs throughout all of their Fulfillment Stories.

Reflexivity
A two-way feedback loop in which perceptions affect an environment and the environment affects perceptions. In a mimetic rivalry, neither party can act without affecting the perceptions and desires of the other.

Romantic Lie
The idea that our choices are completely autonomous, independent, and self-directed. Someone under the power of the Romantic Lie never thinks of their behavior as mimetic.

Sacrificial Substitution
Sacrificing something symbolic as a stand-in for another (usually more violent) sacrifice.

Scapegoat
A person, group, or thing that a community chooses to expel or eliminate in the midst of a mimetic crisis in order to bring about a resolution. The scapegoat absorbs all of the mimetic tension and violence, which had previously been undirected and chaotic, onto itself. The scapegoat is often chosen randomly, through a mimetically driven process of judgment.

Scapegoat mechanism
The process by which humans have historically resolved mimetic crises by expelling or otherwise eliminating a scapegoat. The first time the scapegoat mechanism is employed, it happens mimetically and spontaneously. After that, it is reenacted in ritual fashion in a way that re-creates and resolves the original crisis, providing temporary catharsis for those involved.

Thick desire*
Thick desires are less mimetic than thin desires. They have had time to form and solidify over many years or during a formative experience that is at the core of a person's life. Thick desires have meaning. They are enduring.

Thin desire*
Thin desires are rooted in ephemeral, superficial things. They're fleeting, mimetic desires that dominate most of life when it is lived unintentionally and easily infected by mimetic phenomena.

Transcendent leadership
A leadership approach that views the generation and formation of desire as the first and most important goal of a leader and the primary driver of organizational culture and health.

APPENDIX B:
MIMETIC THEORY READING LIST

I believe that a person's intellectual journey is relatively path dependent. There's a progression of books that I recommend for mimetic theory. With that said, different people should start in different places and progress according to their interests and motivation. Some people will want to jump directly to *Things Hidden Since the Foundation of the World*, arguably Girard's magnum opus. The list below is in rough sequential order and simulates the progression that I would probably use if I were to design a year-long mimetic theory seminar.

1. *Deceit, Desire, and the Novel: Self and Other in Literary Structure*, René Girard (1961)
2. *I See Satan Fall Like Lightning*, René Girard (1999)
3. *René Girard's Mimetic Theory*, Wolfgang Palaver (2013)
4. *Things Hidden Since the Foundation of the World*, René Girard (1978)
5. *Evolution of Desire: A Life of René Girard*, Cynthia L. Haven (2018)
6. *Violence Unveiled: Humanity at the Crossroads*, Gil Bailie (1995)
7. *Mimesis and Science: Empirical Research on Imitation and the Mimetic Theory of Culture and Religion*, Scott R. Garrels, editor (2011)
8. *Evolution and Conversion: Dialogues on the Origins of Culture*, René Girard (2000)
9. *Resurrection from the Underground: Feodor Dostoevsky*, René Girard (1989)

10. *Battling to the End: Conversations with Benoît Chantre*, René Girard (2009)

To continue the discussion, visit lukeburgis.com and follow Luke on Twitter @lukeburgis.

APPENDIX C:
MOTIVATIONAL THEMES

The following are themes of the twenty-seven motivational patterns identified in the System for Identifying Motivated Abilities (SIMA). The MCODE (Motivation Code) is an online assessment that draws on the underlying discoveries of SIMA. It uses a narrative, storytelling process and takes around forty-five minutes to complete.

If you'd like to take the assessment, visit lukeburgis.com/motivation for guidance and a reader discount.

Achieve Potential: Identifying and realizing potential is a constant focus of your activities.

Advance: You love the experience of making progress as you accomplish a series of goals.

Be Unique: You seek to distinguish yourself by displaying some talent, quality, or aspect that is distinctive and special.

Be Central: You are motivated to be a key person who holds things together and gives them meaning and/or direction.

Bring Control: You want to be in charge and in control of your own destiny.

Bring to Completion: Your motivation is satisfied when you can look at a finished product or final result and know that your work is done and that you have met the objective you set out to accomplish.

Comprehend and Express: Your motivation focuses on understanding, defining, and then communicating your insights.

Collaborate: You enjoy being involved in efforts in which people work together for a common purpose.

Demonstrate New Learning: You are motivated to learn how to do something new and show that you can do it.

Develop: You are motivated by the process of building and developing from start to finish.

Evoke Recognition: You are motivated to capture the interest and attention of others.

Experience the Ideal: You are motivated to give concrete expression to certain concepts, visions, or values that are important to you.

Establish: You are motivated to lay secure foundations and to be established.

Explore: Pressing beyond the existing limits of your knowledge and/or experience, you explore what is unknown or mysterious to you.

Excel: You want to excel or at least to do your absolute best as you exceed the performance or expectations of those around you.

Gain Ownership: The nature of your motivation is expressed through efforts to acquire what you want and to exercise ownership or control over what is yours.

Improve: You are happiest when you are using your abilities to make things better.

Influence Behavior: You are motivated to gain a reaction or response from people that indicates you have influenced their thinking, feelings, and behavior.

Make an Impact: You seek to make an impact or personal mark upon the world around you.

Make It Right: You consistently set up or follow standards, procedures, and principles that you believe are "right."

Make It Work: Your motivation focuses on fixing something that has broken down or is functioning poorly.

Make the Grade: You are motivated to make the grade and gain acceptance into a group in which you want to be a member or participant.

Master: Your motivation is satisfied when you are able to gain complete command of a skill, subject, procedure, technique, or process.

Meet the Challenge: Your sense of achievement comes in looking back over a challenge you have met or a test you have passed.

Organize: You want to set up and maintain a smooth-running operation.

Overcome: Your motivation focuses on overcoming and winning out over difficulties, disadvantages, or opposition.

Serve: You are motivated to identify and fulfill needs, requirements, and expectations.

What follows is a sample interview and partial results from an MCODE assessment. I asked Maria (not her real name), an employee at a digital marketing firm, a few simple questions to draw out her Fulfillment Stories. This is an adaptation of our exchange:

> **Luke**: "Can you tell me about a time in your life when you took an action that you believe you did well and which brought you a sense of fulfillment? It can come from any time in your life. It doesn't matter if you were seven years old or thirty-seven."
>
> **Maria**: "In my senior year of cross country running I was able to finish out my career with a strong season by placing in the top three in my last meet, which allowed me to go compete in regionals in New England."
>
> **Luke**: "Can you tell me what you did—what specific actions you took—to accomplish that?"
>
> **Maria**: "I got into the best shape of my life by following a strict diet, getting up at 5 a.m. to train, and limiting the kind of social invites that I accepted so that I didn't get off track in my preparation."
>
> **Luke**: "And what specifically about that accomplishment was the most fulfilling thing for you?"
>
> **Maria**: "I was able to earn the respect of my coach and teammates. Before the meet, I think they had only ever seen me as average. I loved traveling to New England for the regional meet with the team, too."

In her last answer, she described what it was, specifically, that gave her the most fulfillment: earning the respect of her coach and teammates. Maria derived satisfaction from following through on her training program, from running well, and from traveling to New England with her teammates. But none of those things were most important to her. Earning the respect of her coach and teammates was what provided her with the most fulfillment. That's what she was after.

In order to get a fuller picture of how Maria is truly motivated, though,

I need to go deeper in order to find a *pattern* in her core motivational drive. I ask her to share two more stories from different periods in her life. Here are Maria's answers, which I've put into tables to show how I like to organize the information, followed by her top three core motivational themes with detailed descriptions.

STORY TWO

THE ACHIEVEMENT	WHAT I ACTUALLY DID	SOURCE OF SATISFACTION
I worked with my husband to execute paying off our student loans.	I figured out a system of sacrifice, budgeting, and hard work and then worked with my husband to pool our resources and pay off the debt in a short period of time. We came together as a team and I had to get very creative to make this happen.	It brought me a sense of freedom in my life and for our future. I felt like this was something that most people would not be able to do in the same set of circumstances.

STORY THREE

THE ACHIEVEMENT	WHAT I ACTUALLY DID	SOURCE OF SATISFACTION
I trained and competed in a marathon.	Not long after having my second baby, I decided to train for a full marathon. I had to get back into shape and go from hardly running at all to running long distances. I also beat the goal time that I had set.	I felt strong and was able to do something like this at a time in my life that I was supposed to be weak. I felt especially proud because I didn't just finish the race but competed at a high level while doing it.

TOP THREE MOTIVATIONAL THEMES

What follows is taken directly from Maria's MCODE (Motivation Code) results. They are general descriptions of her top three motivational themes. Do any of them sound like yours?

1. EXCEL

You want to excel or, at the very least, do your absolute best as you exceed the performance or expectations of those around you. You thrive on competition. Maybe you compete against yourself in efforts that test your limits and stimulate you to stretch yourself to develop your skills, understanding, or expertise to the greatest extent possible. Certain standards of excellence, efficiency, or quality might be the main focus of your competitive drive. It may be that competing head-to-head with other people might be what you enjoy most. In any case, you identify challenges that give you a clear shot at exceeding your own previous efforts, the efforts of others, or the typical performance. With the goal clearly in mind, you focus your abilities on an effort to excel. Achievement for you means surpassing the requirements of your work, responsibility, or position. You want to establish a reputation that confirms the excellence of your work. In general, you want to do better than others—for example, to be the fastest or most effective.

2. OVERCOME

Your motivation focuses on overcoming and winning out over difficulties, disadvantages, or opposition. Determination, persistence, and a competitive spirit tend to be among your natural traits. You enjoy exerting a sustained, all-out effort to prevail over problems, difficulties, obstacles, handicaps, or adversaries. Your narrative may feature achievements like getting your degree while working full-time and supporting a family. Despite a painful injury, you might gut it out and perform well in an athletic contest. You might strive to handle demanding job responsibilities despite a lack of experience, inadequate skills, or inadequate educational background. There may be a story where you work to prove the validity of an idea or plan of yours that others have scoffed at. You may have politicked heavily to overcome opposition to one of your proposals. You are motivated to struggle with the factors working against you until you overcome them.

3. MASTER

Your motivation is satisfied when you are able to gain complete command of a skill, subject, procedure, technique, or process. You want your knowledge, execution, or control over the intricacies and details involved to be flawless. The focus of your attention may be a sales technique, production procedure, or a core method employed in a trade or craft. You might want to master a sport such as golf, tennis, or skiing. It could be that you concentrate on the principles behind an engineering problem or on some economic, scientific, or philosophical concept. You may seek command over a system, the variables of a process, or the various elements of a multifaceted job responsibility. It may be some element of your character or your nature that you seek to perfect. In any case, your achievements are full of such examples. Your thinking and talents are oriented toward mastery, your goals toward perfection.

NOTES AND SOURCES

NOTES

Note to Reader

1. "Peter Thiel on René Girard," ImitatioVideo, YouTube, 2011. https://www.youtube.com/watch?v=esk7W9Jowtc.

Prologue

1. Tony Hsieh, *Delivering Happiness: A Path to Profits, Passion and Purpose*, 191, Grand Central Publishing, 2010.
2. I am using the phrase "skin in the game" in the spirit of Nassim Nicholas Taleb and his excellent book by that name. At that point in my life, I was what Taleb would call fragile—the amount of debt that I had constrained my options. Worst of all, my desires were fragile.
3. I learned this from the psychologist Jean-Michel Oughourlian, a close friend of Girard's, who likes to describe mimetic desire as a movement of desire that draws people together—and then pushes them apart.

Introduction

1. Peter Thiel and Blake Masters, *Zero to One: Notes on Startups, or How to Build the Future*, Crown Business, 2014.
2. Paul J. Nuechterlein, "René Girard: The Anthropology of the Cross as Alternative to Post-Modern Literary Criticism," *Girardian Lectionary*, October 2002.
3. Girard uses the word "desire" (or *désir* in French) because desire was a hotly debated category in philosophical circles in mid-twentieth-century France. After World War II, the question of "desire" dominated French literature and intellectual life. When Girard began exploring the topic, Sigmund Freud, Jean-Paul Sartre, Alexandre Kojève, Jacques Derrida, and others were already wrestling with it. So Girard took up their category (*désir*) and

radically transformed it. For Girard, desire is the most salient feature of the human condition and imitation the most fundamental feature of human behavior.

4. Mimetic desire may be something that sociologist Émile Durkheim, if he were alive today, would call a *social fact*. In *The Rules of Sociological Method* (Oxford Reference, 1895, 1964) Durkheim describes a social fact as an aspect of social life that shapes or constrains a person's action.

5. James Alison, *The Joy of Being Wrong: Original Sin Through Easter Eyes*, Crossroad, 1998.

6. Sandor Goodhart, "In Tribute: René Girard, 1923–2015," *Religious Studies News*, December 21, 2015.

7. René Girard, *Conversations with René Girard: Prophet of Envy*, ed. Cynthia Haven, Bloomsbury, 2020.

8. Cynthia Haven, *Evolution of Desire: A Life of René Girard*, Michigan State University Press, 2018.

9. Haven, *Evolution of Desire*, 288.

10. *Apostrophes*, episode 150, France 2, June 6, 1978.

11. René Girard, Jean-Michel Oughourlian, and Guy Lefort, *Things Hidden Since the Foundation of the World*, Stanford University Press, 1987.

12. Thiel and Masters, *Zero to One*, 41.

13. Trevor Cribben Merrill, *The Book of Imitation and Desire: Reading Milan Kundera with René Girard*, Bloomsbury, 2014.

14. René Girard and Benoît Chantre, *Battling to the End: Conversations with Benoît Chantre*, 212, Michigan State University Press, 2009.

Chapter 1: Hidden Models

1. James Warren, *Compassion or Apocalypse: A Comprehensible Guide to the Thought of René Girard*, Christian Alternative, 2013.

2. Jean-Michel Oughourlian, *The Genesis of Desire*, Michigan State University Press, 2010.

3. Francys Subiaul, "What's Special About Human Imitation? A Comparison with Enculturated Apes," *Behavioral Sciences* 6, no. 3, 2016.

4. Sophie Hardach, "Do Babies Cry in Different Languages?," *New York Times*, November 14, 2019. Also: Birgit Mampe, Angela D. Friederici, Anne Christophe, and Kathleen Wermke, "Newborns' Cry Melody Is Shaped by Their Native Language," *Current Biology* 19, no. 23, 2009.

5. Adapted from the first paragraph of Andrew Meltzoff's essay, "Out of the Mouths of Babes: Imitation, Gaze, and Intentions in Infant Research—the 'Like Me' Framework," in *Mimesis and Science: Empirical Research on Imitation and the Mimetic Theory of Culture and Religion*, ed. Scott R. Garrels, Michigan State University Press, 2011.

6. A. N. Meltzoff and M. K. Moore, "Newborn Infants Imitate Adult Facial Gestures," *Child Development* 54, 1983, 702–09. Photo credit: A. N. Meltzoff and M. K. Moore, "Newborn Infants Imitate Adult Facial Gestures," *Science* 198, 1977, 75–78.

7. A. N. Meltzoff, "Out of the Mouths of Babes," in *Mimesis and Science*, 70.

8. Marcel Proust, *In Search of Lost Time*, vol. 5, *The Captive, The Fugitive*, 113, Modern Library edition, Random House, 1993.

9. A. N. Meltzoff, "Understanding the Intentions of Others: Re-enactment of Intended Acts by 18-Month-Old Children," *Developmental Psychology* 31, no. 5, 1995, 838–50.

10. Rodolfo Cortes Barragan, Rechele Brooks, and Andrew Meltzoff, "Altruistic Food Sharing Behavior by Human Infants After a Hunger Manipulation," *Nature Research*, February 2020.

11. A. N. Meltzoff, R. R. Ramírez, J. N. Saby, E. Larson, S. Taulu, and P. J. Marshall, "Infant Brain Responses to Felt and Observed Touch of Hands and Feet: A MEG Study," *Developmental Science* 21, 2018, e12651.

12. Eric Jaffe, "Mirror Neurons: How We Reflect on Behavior," *Observer*, May 2007.

13. Sue Shellenbarger, "Use Mirroring to Connect with Others," *Wall Street Journal*, September 20, 2016.

14. Larry Tye, *The Father of Spin: Edward L. Bernays and the Birth of Public Relations*, Henry Holt, 2002.

15. Adam Curtis, director, *The Century of the Self*, BBC Two, March 2002.

16. From the documentary film *The Century of the Self*.

17. Tye, *The Father of Spin*, 23.

18. Nobody has done more than Martha J. Reineke, professor in the Department of Philosophy and World Religions at the University of Northern Iowa, to explore gender and femininity through the lens of mimetic theory. Her work with mimetic theory is wide-ranging, but it is especially pertinent and a critical contribution to women's studies, which regrettably has not received the same level of attention that other applications of Girard's thought have received.

19. Tye, *The Father of Spin*, 30.

20. Podcast *Entitled Opinions* with Robert Harrison on the website stanford.edu. See episode from September 17, 2005, "René Girard: Why We Want What We Want." https://entitledopinions.stanford.edu/ren-girard-why-we-want-what-we-want.

21. From the story told on the *Entitled Opinions* podcast, September 17, 2005, beginning around the 15:00 mark.

22. Adam M. Grant, *Give and Take*, 1–3, Viking, 2013.

23. David Foster Wallace, "E Unibus Pluram: Television and U.S. Fiction," *Review of Contemporary Fiction* 13, no. 2, Summer 1993, 178–79. Earlier in this work, on page 152, he writes: "Television, from the surface on down, is about *desire*. Fictionally speaking, desire is the sugar in human food."

24. The *BBC Business Daily* podcast, "Tesla: To Infinity and Beyond," February 12, 2020.

25. The information theory of economics draws on the work of mathematician Claude Shannon, founder of information theory, to show the central importance of information in the economy and the forces that work to suppress and support the flow of information and the impact this has on value creation. George Gilder has defended capitalism through the lens of information theory in his book *Knowledge and Power: The Information Theory of Capitalism and How It Is Revolutionizing Our World*. In my view, information theory is incomplete on its own due to the limited role that information plays in a robust human ecology. It needs to be supplemented with an understanding of the role of mimetic desire, among other things. Mathematician Benoit B. Mandelbrot made a case for the irrationality of the market and the silliness of traditional financial theories in his book, with Richard L. Hudson, *The Misbehavior of Markets: A Fractal View of Financial Turbulence*, Basic Books, 2006.

26. Jason Zweig, "From 1720 to Tesla, FOMO Never Sleeps," *Wall Street Journal*, June 17, 2020.

Chapter 2: Distorted Reality

1. Yalman Onaran and John Helyar, "Fuld Solicited Buffett Offer CEO Could Refuse as Lehman Fizzled," *Bloomberg*, 2008.

2. Walter Isaacson, *Steve Jobs*, Simon & Schuster, 2011.

3. The word "mesmerize" comes from the nineteenth-century Viennese physician Franz Mesmer—the father of what we now call hypnosis. Mesmer believed there was a force that drew certain people to other people and things. He was one of the first people to start thinking that there may be laws of motion for psychic or social realities that were analogous to Newton's laws of physics. "One must grant Newton the greatest praise," he wrote, "because he has clarified to the highest degree the reciprocal attraction of all things" (Oughourlian, *The Genesis of Desire*, 84). Mesmer went on to conduct individual and group therapy sessions in which he'd manipulate patients with his hands and finish off sessions by playing a glass harmonica. Many of his patients began to report miraculous healings. You could say that they were subject to a placebo effect. But that would be to underestimate the relationship between patient and doctor—the magnetism of Mesmer himself and the suggestibility of his patients when under his influence. Modern hypnosis works based on the power of suggestion that the hypnotist wields over the subject. The hypnotist is a model of desire—what the hypnotist wants, the subject wants. It's no surprise that the longest-running and most successful hypnotist show of all time, the female hypnotist Pat Collins, took place at a club in Hollywood. The people who showed up to her shows were the kinds of people who desperately wanted to "make it," and they were more open to the suggestions of a mediator than most.

4. Michael Balter, "Strongest Evidence of Animal Culture Seen in Monkeys and Whales," *Science Magazine*, April 2013.

5. *Models* and *mediators* are the same thing. Mediation is what models do—they give their subjects new eyes to see and value things in a new light.

6. Tobias Huber and Byrne Hobart, "Manias and Mimesis: Applying René Girard's Mimetic Theory to Financial Bubbles," *SSRN*, 24.

7. See René Girard, *Deceit, Desire, and the Novel*, 53–82, trans. Yvonne Freccero, Johns Hopkins University Press, 1976. I have wondered why he didn't use the term "ontological desire," which seems to get more directly at the idea of a desire for being itself. I have not found a good explanation of this word choice. However, I do believe that we can understand Girard's take on metaphysical desire easily if we think of it as "after the physical"—when we have metaphysical desire, we can't be satisfied by any physical objects. "As the role of metaphysical grows greater in desire, that of the physical diminishes in importance," writes Girard (85). Contrast this with my dog, who lies down and goes to sleep after I feed him a good meal. He doesn't stare up at the stars, howling, wondering what comes next.

8. Today, "metaphysics" usually refers to first principles, the things that form the foundation from which all other things spring. Elon Musk claims to use them as the key to his decision-making process. Tim Higgins of the *Wall Street Journal* wrote: "The billionaire has attributed his business success to the scientific approach called First Principles, which is rooted in Aristotle's writings and among other things rejects solving problems with copycat solutions and depends upon reducing problems to their essence even if the solutions might seem counterintuitive." Tim Higgins, "Elon Musk's Defiance in the Time of Coronavirus," *Wall Street Journal*, March 20, 2020.

9. René Girard and Mark Rogin Anspach, *Oedipus Unbound*, 1, Stanford University Press, 2004.

10. René Girard, *Anorexia and Mimetic Desire*, Michigan State University Press, 2013.

11. THR Staff, "Fortnite, Twitch . . . Will Smith? 10 Digital Players Disrupting Traditional Hollywood," *Hollywood Reporter*, November 2018.

12. René Girard, *Resurrection from the Underground: Feodor Dostoevsky*, trans. James G. Williams, Michigan State University Press, 2012.

13. Virginia Woolf, *Orlando*, Edhasa, 2002.

14. House of Lords, October 28, 1943. The House of Commons had been bombed in May 1941, and Churchill demanded it be rebuilt exactly as it was before. Quote found in Randal O'Toole's "The Best-Laid Plans," 161, Cato Institute, 2007.

15. Negative imitation is related to *negative partisanship*, in which people form their political ideas based on what another party's ideas are. Girard, *Deceit, Desire, and the Novel*.

16. The most mimetic episodes of *Seinfeld* are "The Soul Mate" (Season 8, Episode 2) and "The Parking Space" (Season 3, Episode 22). After you've finished this chapter, I recommend pouring yourself a glass of wine and watching them for a hilarious illustration of mimetic desire and mimetic rivalry at work. Commenting on *Seinfeld* in general, Girard writes: "In order to be successful an artist must come as close as he can to some important social truth without inciting painful self-criticism in the spectators. This is what this show did. People do not have to understand fully in order to appreciate. They must not understand. They identify themselves with these characters because they do it, too. They recognize something that is common and true, but they cannot define it. Probably the contemporaries of Shakespeare appreciated his portrayal of human relations in the same way we enjoy *Seinfeld*, without understanding his perspicaciousness about mimetic interaction. I must say that there is more social reality in *Seinfeld* than in most academic sociology." (René Girard, *Evolution and Conversion: Dialogues on the Origins of Culture*, 179, Bloomsbury, 2017.)

17. Dubious origins. It seems as if Marx wrote something like this in a resignation letter to a club he belonged to.

Chapter 3: Social Contagion

1. Tribune Media Wire, "Man in Coma After Dispute over Towel Sparks Massive Brawl at California Water Park," *Fox31 Denver*, August 26, 2019.

2. My source for most of the material on Ferruccio Lamborghini is a rare book written by Tonino Lamborghini, son of Ferruccio, which I found in Italy (where I lived for three years between 2013 and 2016). It is available only in Italian. All translations of dialogue taken from the story are my own. The book is *Ferruccio Lamborghini: La Storia Ufficiale* by Tonino Lamborghini, Minerva, 2016, a tribute of a son to his father. Of course, the story is told from only one side. When I searched Luca Dal Monte's nearly 1,000-page biography of Enzo Ferrari, *Enzo Ferrari: Power, Politics, and the Making of an Automotive Empire*, David Bull Publishing, 2018, I found that not a single mention of Ferruccio Lamborghini is made. A conspicuous omission—or evidence of things hidden since the foundation. . . .

3. Lamborghini, *Ferruccio Lamborghini*.

4. "The Argument Between Lamborghini and Ferrari," WebMotorMuseum.it. https://www.motorwebmuseum.it/en/places/cento/the-argument-between-lamborghini-and-ferrari/.

5. Nick Kurczewski, "Lamborghini Supercars Exist Because of a 10-Lira Tractor Clutch," *Car and Driver*, November 2018.

6. But Lamborghini was motivated by more than rivalry. A move into car manufacturing made business sense. The profit margins of luxury cars dwarfed those for tractors. He also saw a gap in the market for high-performance cars. Nobody had produced a vehicle that matched Ferrari's power on the racetrack but that also had a luxury interior. This would be his niche— the Lamborghini would make his supercar a *granturismo*, or touring car, that matched the power of a Ferrari but exceeded it in comfort.

7. Lamborghini, *Ferruccio Lamborghini*.

8. Austin Kleon, *Steal Like an Artist: 10 Things Nobody Told You About Being Creative*, 8, Workman, 2012.

9. Contrary to popular belief, the bull's charging has nothing to do with his hatred for the color red. Bulls are colorblind; it seems they just get irritated by the waving movement.

10. Girard, *Deceit, Desire, and the Novel*, 176.

11. Lamborghini, *Ferruccio Lamborghini*.

12. Susan Blackmore, an expert of meme theory and author of *The Meme Machine* (Oxford University Press, 2000), is explicit about the role of imitation. I recommend her book to anyone looking for a good introduction to memes.

13. Dawkins's original articulation of meme theory said little about why certain memes are selected for imitation in the first place. He said that memes mutate "by random change and a form of Darwinian selection" (Olivia Solon, "Richard Dawkins on the Internet's Hijacking of the Word 'Meme,'" *Wired UK*, June 20, 2013). In mimetic theory, objects are selected through mimetic selection—that is, they are chosen because a model chose them first.

14. James C. Collins, *Good to Great*, 164, Harper Business, 2001.

15. James C. Collins, *Turning the Flywheel: Why Some Companies Build Momentum . . . and Others Don't*, 9–11, Random House Business Books, 2019.

16. Collins, *Turning the Flywheel*, 11.

17. Aristotle, *Metaphysics*, Book IX (Theta), trans. W. D. Ross, Oxford University Press, rev. ed., 1924. See also http://classics.mit.edu/Aristotle/metaphysics.9.ix.html.

18. In Tony Hsieh's book, *Delivering Happiness: A Path to Profit, Passion, and Purpose* (p. 58), Nick Swinmurn, the original founder of Zappos, is quoted as saying, "Buying a pair of shoes shouldn't be so hard, I remember thinking," to describe how he got his original idea. It wasn't delivering happiness. Tony Hsieh writes (p. 56): "His idea was to build the Amazon of shoes and create the world's largest shoe store online."

19. Tony shares in his book that he sent an email to all Zappos employees in October 2000 stating the importance of focusing on gross profit, increasing new qualified visitors to the website, and increasing the percentage of repeat customers. His email included the following exhortation: "We need everything to think about how it will increase our total gross profit over the next 9 months. This will mean that some projects that we might normally pursue will have to be put on hold until we get to profitability. Once we get to profitability, then we will be able to think longer-term and bigger picture, and fantasize more about how to rule the world."

20. "By the end of lunch, we realized that the biggest vision would be to build the Zappos brand to be about the very best customer service," Tony writes in *Delivering Happiness* (p. 121), recalling a lunch he had with Mossler in early 2003.

21. This was the value of the deal on the day of closing. Because it was an all-stock deal (Zappos received Amazon stock, not cash), the value of the deal is pegged to the value of Amazon stock at any given time. The closing price of Amazon stock on October 30, 2009, was $117.30. As of the time of this writing, it's about $3,423. Zappos was smart to demand an all-stock deal.

22. Nellie Bowles, "The Downtown Project Suicides: Can the Pursuit of Happiness Kill You?," *Vox*, October 1, 2014.

23. "Tony Hsieh's "Rule for Success: Maximize Serendipity," *Inc.com*, January 25, 2013.

24. Brian J. Robertson, *Holacracy: The New Management System for a Rapidly Changing World*, Henry Holt, 2015.

25. To many outsiders, the move to holacracy seemed likely to end in chaos and confusion once people couldn't pull rank on others and didn't have structure to cling to. After all, humans started out with self-organization and used it to build the very hierarchical structures that holacracy wants to throw out. But holacracy was a natural extension of how Tony always liked to operate. Like many others in Silicon Valley, Tony was a Burner—a longtime devotee of the Burning Man festival, a huge gathering held every year in the Black Rock Desert in northwestern Nevada. Burners tend to have a strong anti-hierarchical ethos, and in a similar spirit, Tony and his team imagined transforming downtown Las Vegas into a radically free-wheeling community where everyone felt empowered to follow their bliss.

26. I use this language intentionally in the spirit of Fyodor Dostoevsky's *Notes from Underground*, which, according to René Girard, is primarily concerned with mimetic desire and rivalry. Girard wrote a volume largely dedicated to the work titled *Resurrection from the Underground: Feodor Dostoevsky*. (Note: the name Fyodor can be spelled various ways in English. I've chosen this way.)

27. C. S. Lewis, "The Inner Ring," para. 16, Memorial Lecture at King's College, University of London, 1944. https://www.lewissociety.org/innerring/.

28. Augustine of Hippo wrote in his book *The City of God* that *ordo amoris* (the order of love) is a "brief and true definition of virtue." Knowing how values hang together and when to pursue things under which circumstances and to what degree—and then developing the will to do so—is the work of a lifetime. The twentieth-century philosopher Max Scheler illustrated an influential hierarchy of values and emotions that shows, in part, how all emotions are not equal. They come about as a response to value, and they can be more or less in accord with the truth of that value. If I respond with joy to another's misfortune, then my affective responses (my emotions) are signaling that something is wrong in my hierarchy of alues—or, more deeply, in my order of love. See the work of Dietrich von Hildebrand and his value-response theory of ethics for further exploration. Dietrich von Hildebrand and John F. Crosby, *Ethics*, Hildebrand Project, 2020.

29. Cap tables differ by company, as do even the names I'm using. They depend on how a company defines different classes of stock, creditor claims, and much more.

30. There's no doubt that people arrive at values partly through mimesis. Aristotle talked about virtues—such as courage, patience, honesty, and justice—as things that people learn to desire because their models either possess or desire them, too. We acquire virtues by imitating our models. (Is it any wonder, then, that in a society in which most people don't value classical virtues, there is little appetite for them?)

31. I have provided some free resources for doing this exercise on my website, https://lukeburgis .com.

32. Bailey Schulz and Richard Velotta, "Zappos CEO Tony Hsieh, Champion of Downtown Las Vegas, Retires," *Las Vegas Review-Journal*, August 24, 2020.

33. Aimee Groth, "Five Years In, Tony Hsieh's Downtown Project Is Hardly Any Closer to Being a Real City," *Quartz*, January 4, 2017.

Chapter 4: The Invention of Blame

1. © 2020 Jenny Holzer, member Artists Rights Society (ARS), New York.

2. René Girard, *The One by Whom Scandal Comes*, 8, trans. M. B. DeBevoise, Michigan State University Press, 2014.

3. Girard, *The One by Whom Scandal Comes*, 7.

4. Carl Von Clausewitz, *On War*, 83, ed. and trans. Michael Howard and Peter Paret, Everyman's Library, 1993.

5. René Girard, *Violence and the Sacred*, trans. Patrick Gregory, Johns Hopkins University Press, 1979.

6. Here is one translation of the prayer that was offered over the goat: "O Lord, I have acted iniquitously, trespassed, sinned before Thee: I, my household, and the sons of Aaron—Thy holy ones. O Lord, forgive the iniquities, transgressions, and sins that I, my household, and Aaron's children—Thy holy people—committed before Thee, as is written in the law of Moses, Thy servant, 'for on this day He will forgive you, to cleanse you from all your sins before the Lord; ye shall be clean.'" Isidore Singer and Cyrus Adler, *The Jewish Encyclopedia: A Descriptive Record of the History, Religion, Literature, and Customs of the Jewish People from the Earliest Times to the Present Day*, 367, Funk and Wagnalls, 1902.

7. From Protestant Reformation scholar William Tyndale's English 1530 translation of the Pentateuch. In Latin, the goat was called *caper emissarius*, or emissary goat—the goat that departs. Tyndale originally rendered it [e]*scape goat*, which eventually became *scapegoat*.

8. René Girard, *I See Satan Fall Like Lightning*, Orbis Books, 2001.

9. Todd M. Compton, *Victim of the Muses: Poet as Scapegoat, Warrior and Hero in Greco-Roman and Indo-European Myth and History*, Center for Hellenic Studies, 2006.

10. The idea that one gun in a firing squad was loaded with a blank has been called into question as a myth by some who say that a real bullet makes a gun recoil too much, while a blank does not, so that anybody taking part in a firing squad would know whether they had a blank. However, we *know* that blanks have been used in firing squads due to documented evidence. Whether every member of the firing squad knew they had a blank is not the most interesting thing; the fact that blanks were used at all is.

11. Localized financial crises work the same way. In 1997, the Asian financial crisis began in Thailand, causing the Thai stock market to plunge by more than 75 percent. The crisis spread quickly to other Asian countries but had a minimal impact on the United States.

12. The idea of a party is inspired by Girard's reflections on Dionysian festivals. These festivals in ancient Greece to honor the god Dionysus were assembled with the purpose of re-creating some form of original unity that had been lost in the chaos of mimetic desire. The festivals re-create the movement from unity to disunity and disorder and culminate in a ritual sacrifice, a scapegoat, which restores order by preventing further disorder and internal conflict.

Raymund Schwager, Girard's close collaborator—who was instrumental in the development of Girard's theological thought—mentions in a letter to Girard that he ran across a fascinating book called *Die Mimesis in der Antike* by Hermann Koller, published in 1954. In it, the author reflects on Plato's use of the Greek word μιμεῖσθαι (*mimesthai*) and concludes that it came from sacred dance. The word "mimos," he writes, designated the actors in the Dionysian festivals.

13. Ta-Nehisi Coates, "The Cancellation of Colin Kaepernick," *New York Times*, November 22, 2019.

14. Girard, *I See Satan Fall Like Lightning*.

15. Flavius Philostratus, *The Life of Apollonius of Tyana, the Epistles of Apollonius and the Treatise of Eusebius*, trans. F. C. Conybeare, Loeb Classical Library, 2 vols., Harvard University Press, 1912.

16. Here is the full passage: "During his illness he'd dreamt that the whole world was condemned to fall victim to some terrible, previously unknown pestilence, which was moving toward Europe out of the depths of Asia. Everyone would perish except for a chosen few, very few. Some kind of new trichina had appeared, and the microscopic organisms settled in human bodies. But these organisms were creatures endowed with intelligence and will. People who were affected immediately became possessed and insane. But never, never did these people consider themselves so intelligent and so infallible about the truth as when they were infected. Never did they consider their pronouncements, their scientific conclusions, their moral convictions and beliefs so infallible. Whole populations, whole towns and nations became infected and went insane. Everyone was anxious, no one understood anyone else, each one thought that truth resided in him alone and, regarding all the others, suffered, beat his chest, wept, and wrung his hands. They didn't know whom to try and how to judge; they couldn't agree on what constituted good and evil. They didn't know whom to condemn and whom to acquit. People killed each other in a senseless rage. They assembled whole armies against one another, but when those armies were on the march, the troops suddenly began to fight among themselves, the ranks disintegrated, the soldiers fell on one another, stabbed and slashed one another, bit and ate one another. In towns the alarm bell sounded all day long: they summoned everyone, but no one knew who had been called or why they had been called, and everyone was anxious. They forsook the most ordinary trades because everyone proposed his own ideas and suggestions, and they were unable to agree; agriculture was abandoned. In some places people formed into groups, agreed on something together, and swore not to disband—but immediately they began to do something quite different from what they had just proposed. They began to accuse one another, to fight and slaughter one another. Conflagrations arose, famine followed. Nearly everything and everyone perished. The pestilence grew and advanced further and further. Only a few people in the whole world could be saved; these were the pure and chosen, destined to found a new race of people and a new life, to renew and purify the earth; but no one had ever seen these people, no one had ever heard their words or their voices." Fyodor Dostoyevsky, *Crime and Punishment*, 600, trans. Michael R. Katz, W. W. Norton, 2019.

17. It's worth exploring the concept of interdividual psychology, a phrase coined by René Girard, Jean-Michel Oughourlian, and Guy Lefort in *Things Hidden Since the Foundation of the World* (Stanford University Press, 1987) to move away from a monadic view of the subject and to properly account for the relationship structure of psychology.

18. Elias Canetti, *Crowds and Power*, 15, Farrar, Straus and Giroux, 1984.

19. From Sophocles's play *Oedipus Rex*, first performed around 429 BCE.

20. Girard, *Violence and the Sacred*, 79.

21. Christian Borch, in his *Social Avalanche: Crowds, Cities, and Financial Markets* (Cambridge University Press, 2020), uses this word to describe crowd psychology.

22. Yun Li, "'Hell Is Coming'—Bill Ackman Has Dire Warning for Trump, CEOs if Drastic Measures Aren't Taken Now," CNBC, March 18, 2020.

23. John Waller, *The Dancing Plague: The Strange, True Story of an Extraordinary Illness*, 1, Sourcebooks, 2009.

24. Ernesto De Martino and Dorothy Louise Zinn, *The Land of Remorse: A Study of Southern Italian Tarantism*, Free Association Books, 2005.

25. Rui Fan, Jichang Zhao, Yan Chen, and Ke Xu, "Anger Is More Influential Than Joy: Sentiment Correlation in Weibo," *PLOS ONE*, October 2014.

26. Stephen King, *On Writing: A Memoir of the Craft*, 76, Scribner, 2010. See also Stephen King, "Stephen King: How I Wrote Carrie," *Guardian*, April 4, 2014, para. 6.

27. *The Hunger Games* series is a contemporary twist on ancient Rome's "bread and circuses." The Romans knew they needed to give the Roman people bread—food to eat—in order to placate them. But they also had to provide circuses, or entertainment. The ritual sacrifice of gladiators or animals protected Rome from its own violence, preventing violent uprisings and keeping its leaders safe.

28. René Girard, *The Scapegoat*, 113, Johns Hopkins University Press, 1996.

29. René Girard and Chantre Benoît, *Battling to the End: Conversations with Benoît Chantre*, xiv, Michigan State University Press, 2009.

30. Girard, *Violence and the Sacred*, 33.

31. John 11:49–50. From the New Revised Standard Version Bible (NRSV), copyright © 1989 the Division of Christian Education of the National Council of the Churches of Christ in the United States of America. Used by permission. All rights reserved.

32. Steven Pinker, in his book *The Better Angels of Our Nature: Why Violence Has Declined* (Penguin Publishing, 2012) debunks what he calls the "hydraulic theory" of violence—the idea that pressure builds up under the surface and needs to be released periodically in violence. To be clear, that is not Girard's theory. The scapegoat mechanism occurs during a mimetic crisis because a group acts practically and strategically—the scapegoat mechanism is a social strategy for defusing violence. While Pinker doesn't specifically mention Girard or the scapegoat mechanism, he touches on the strategic nature of violence: "When a tendency toward violence evolves, it is always strategic. Organisms are selected to deploy violence only in circumstances where the expected benefits outweigh the expected costs" (p. 33).

33. This show appeared on Canadian television in March 2011 in David Cayley's Classic IDEAS series.

34. I am not as familiar with texts from other religious traditions, like Buddhism or Hinduism or Islam, but I would welcome a further discussion about how those texts, too, might reveal the scapegoat mechanism, and I invite any scholars of these texts to join the subreddit r/MimeticDesire.

35. Girard, *I See Satan Fall Like Lightning*.

36. René Girard called these *texts of persecution*. They are texts written by the persecutors and they cover up a crime or obscure the truth about what happened. Girard, *The Scapegoat*.

37. Girard articulates this viewpoint most forcefully in his books *I See Satan Fall Like Lightning* and *Evolution and Conversion: Dialogues on the Origins of Culture* (Bloomsbury, 2017).

38. Girard, *I See Satan Fall Like Lightning*, 161.

39. The first hospital is often considered to be the one built by Saint Basil of Caesarea outside the city of Caesarea, modern-day Kayseri, Turkey.

40. Girard, *I See Satan Fall Like Lightning*, foreword, xxii–xxiii.

41. Jesus calls them to task for this hypocrisy in an encounter with some Pharisees: "And you say, 'If we had lived in the days of our ancestors, we would not have taken part with them in shedding the blood of the prophets'" (Matthew 23:30).

42. Aleksandr Solzhenitsyn, *The Gulag Archipelago 1918–1956*, HarperCollins, 1974, 168.

43. Ursula K. Le Guin, *The Ones Who Walk Away from Omelas: A Story*, 262, Harper Perennial, 2017. Excerpted from *The Wind's Twelve Quarters*, originally published in hardcover in 1975 by HarperCollins.

44. René Girard, *The Scapegoat*, 41, Johns Hopkins University Press, 1996.

Part II: The Transformation of Desire

1. David Lipsky, *Although of Course You End Up Becoming Yourself: A Road Trip with David Foster Wallace*, 86, Broadway Books, 2010.

Chapter 5: Anti-Mimetic

1. James Clear, *Atomic Habits: An Easy and Proven Way to Build Good Habits and Break Bad Ones*, 27, Random House Business, 2019.

2. George T. Doran, "There's a S.M.A.R.T. Way to Write Management's Goals and Objectives," *Management Review*, November 1981.

3. Donald Sull and Charles Sull, "With Goals, FAST Beats SMART," *MIT Sloan Management Review*, June 5, 2018.

4. John Doerr, *Measure What Matters: How Google, Bono, and the Gates Foundation Rock the World with OKRs*, Penguin, 2018.

5. Lisa D. Ordóñez, Maurice E. Schweitzer, Adam D. Galinsky, and Max H. Bazerman, *Goals Gone Wild: The Systematic Side Effects of Over-Prescribing Goal Setting*, Harvard Business School, 2009.

6. Sociologist Max Weber called the rigid structure within which many people in organizations make decisions an "iron cage." The fancy way of saying this is "institutional isomorphism." The term was coined by Paul J. DiMaggio and Walter W. Powell in their paper "The Iron Cage Revisited: Institutional Isomorphism and Collective Rationality in Organizational Fields," *American Sociological Review* 48, no. 2, 1983, 147–60, where they describe the mimetic processes that lead to it. This structure imprisons decision-making in a "rationalistic framework," according to Weber, which, barring a revolution, endures "until the last ton of fossilized coal is burnt." The iron cage is not rationalistic, it's mimetic. For further reading, see Max Weber, *The Protestant Work Ethic and the Spirit of Capitalism*, Merchant Books, 2013.

7. Eric Weinstein, interview with Peter Thiel, *The Portal*, podcast audio, July 17, 2019.

8. Mark Granovetter, "Economic Action and Social Structure: The Problem of Embeddedness," *American Journal of Sociology*, November 1985, 481–510.

9. Gargillou includes fern, amaranth, white borage, rocambole garlic, clover, cauliflower stalk, peas, tuberous chervil, nasturtium, phyteuma, pattypan squash, Welsh onion (*Allium fistulosum*), endive, chickweed (*Stellaria media*), pink radish, salsify, tomato, spring onion, Alpine

fennel, and many other vegetables, young shoots, leaves, stalks, grains, or roots depending on the season and even the day.

10. Orson Scott Card, *Unaccompanied Sonata*, Pulphouse, 1992.

11. Mark Lewis, "Marco Pierre White on Why He's Back Behind the Stove for TV's Hell's Kitchen," *The Caterer*, April 2007.

12. Marc Andreessen, "It's Time to Build," Andreessen Horowitz. https://a16oz.com/2020/04/18 /its-time-to-build/.

13. Many exemplary profiles of Michelin as a company are included in the volume *And Why Not? The Human Person and the Heart of Business* by François Michelin (Lexington Books, 2003). Among my favorites is this story of how a young François learned a lesson from his grandfather, one of the founders of the company, about the role of empathy in breaking the cycle of scapegoating: "I remember a day in the year 1936 when, while I was sitting with my grandfather in his office on Cours Sablon, a long line of people on strike passed beneath our windows. When I heard some noise, I went to the window and lifted up the curtain, which prompted shouting. My grandfather said to me: 'People will tell you that these people are nasty, but that is not true.' I realized that my grandfather was speaking the truth, which leads me to say that the notion of class struggle stems from an intellectual laziness in which people want to avoid asking themselves real questions. Since that time, I have been haunted by this phrase of his: 'If you regard a Communist as a class enemy, you are making a mistake. If you see him as a man who simply has a way of thinking that is different from yours, that is a totally different matter.' Each time I meet someone, I ask myself: *What diamond is hidden in this person?* All these jewels that surround us make an incredible crown, when we have learned how to open our eyes and see them" (p. 66).

14. Posted on the BRAS official Facebook page on September 20, 2017. You can watch it at wanting.ly/bras.

15. Resentment (*ressentiment*, in French, which has a more nuanced meaning—implying that one's values or worldview become deeply deformed through resentment) is a phenomenon grappled with by the philosophers Friedrich Nietzsche and Max Scheler. Neither one of them noticed the role of internal mediation in resentment like Girard.

16. Le Suquet still had two stars in the 2020 guide at the time of this writing.

Chapter 6: Disruptive Empathy

1. "Disruptive Empathy" is the title of a section in Gil Bailie's book, *Violence Unveiled: Humanity at the Crossroads*, Crossroad, 2004.

2. René Girard, *The One by Whom Scandal Comes*, 8, trans. M. B. DeBevoise, Michigan State University Press, 2014.

3. Thomas Merton, *New Seeds of Contemplation*, 38, New Directions Books, 2007.

4. René Girard, Robert Pogue Harrison, and Cynthia Haven, "Shakespeare: Mimesis and Desire," *Standpoint*, March 12, 2018.

5. During World War II, Allied planes landed on islands in the south Pacific Ocean for stopovers during long missions. The American and European soldiers gave the locals a handsome bounty of food and sundries in exchange for their goodwill. They'd deliver the goods by parachute by dropping large crates out of cargo planes.

The locals—many of whom fished with spears and lived in huts—were greatly amused:

cigarettes, beef sticks, T-shirts, whiskey, playing cards, handkerchiefs, and propane lights were like talismans from another civilization. We can only imagine the whiskey-fueled conversations around the firepit on nights when a new supply of goods was dropped.

Then the war ended. The locals were stricken. Why, all of a sudden, did the goods they had been receiving suddenly stop appearing? Within a few months, locals began gathering on the runways where the planes had landed. They mimicked air traffic controller motions, carved wooden headphones, and set up makeshift control towers. They lit signal fires and formed parade formations in imitation of the ones they saw the soldiers make after landing. They were imitating the actions they had seen in the hopes of producing the same result.

"They're doing everything right," said Richard Feynman, a Nobel Prize–winning physicist in his commencement speech at Caltech in 1974. "It looks exactly the way it looked before. But it doesn't work . . . they follow all the apparent precepts and forms, but they're missing something essential because the planes don't land." In his speech about science, pseudoscience, and how not to fool yourself, he coined the controversial term "cargo cults" to describe what happened in the South Pacific islands.

The name is misleading for many reasons. For example: the cults are clearly not about cargo at all. We know that because the cults took different forms on the Pacific islands. One businessman who worked on large engineering projects on Lihir Island, off the coast of Papua New Guinea, in the late 1970s, remembers concession stands set up at the airstrip with locals standing around playing the role of businessmen. It was important for each of them to play the role of a businessman, but nobody bought or sold anything. Rather than imitating the cargo plane landings, they were imitating the businessmen who came to scope out new projects to embark on.

The imitation is not primarily for the purpose of making merchandise fall from the sky. It's not even for the purpose of material goods at all. The imitation happens because people want a certain amount of status and respect from others. By imitating those who already seem to have it, the imitators hope that some kind of transformation will take place.

The worst mistake is thinking of this phenomenon as something that "primitive" people do. The cult is a *cult of imitation*—and it's universal. The cargo cults that were publicized in the decades following the war are only the extreme manifestations of what happens every day in the United States and abroad.

Young college graduates (or college dropouts) dress in jeans and T-shirts, work in incubator spaces with brand stickers on the backs of their MacBook Pros, make corporate cultures feel like fraternities (complete with ping-pong tables, kombucha, and craft beer on tap), follow Gary Vaynerchuk on social media, and meet with other cult members in the evening at hip inner-city cafés—all in the hopes that it will increase their company's valuation.

Today, saying "I want to be an entrepreneur" is the equivalent of saying "I want to be a consultant or investment banker" in the early 1990s. But "being an entrepreneur" is a particularly problematic category because entrepreneurship always needs to be concretized around a specific problem or need in the world. To say that one wants to be an entrepreneur even before understanding a unique and particular opportunity is like walking around with a giant mallet looking for things to break. To a man with a hammer, everything looks like a nail. To someone who wants to be an entrepreneur, everything looks like an opportunity to start a company.

6. Here are a few: Why is there something rather than nothing? What is beauty? How do I distinguish between good and evil? What is a conscience? Who am I? Where did I come from? Where am I going?

7. Parker Palmer, *Let Your Life Speak: Listening for the Voice of Vocation*, Jossey-Bass, 1999.

8. Jonathan Sacks, "Introduction to Covenant and Conversation 5776 on Spirituality," October 7, 2015. https://rabbisacks.org.

9. While it's not necessary to go through the formal assessment process in order to get the benefits of the Fulfillment Story exercise, I recommend the tool because it gives insights and language that would be difficult to arrive at on your own. For those interested in learning more, I've included in Appendix C a full example of a person's top three core motivational results along with resources that I use in my class and my company. See Todd Henry, Rod Penner, Todd W. Hall, and Joshua Miller, *The Motivation Code: Discover the Hidden Forces That Drive Your Best Work*, Penguin Random House, 2020.

Chapter 7: Transcendent Leadership

1. Whitney Wolfe Herd in an interview with *CNN Business* on December 13, 2019. Sara Ashley O'Brien, "She Sued Tinder, Founded Bumble, and Now, at 30, Is the CEO of a $3 Billion Dating Empire." https://www.cnn.com/2019/12/13/tech/whitney-wolfe-herd-bumble-risk-takers/index.html.

2. I adapted this language from contemporary philosopher Byung-Chul Han's book *The Burnout Society*. In writing about neuronal illnesses and our inability to develop "antibodies" against them because they don't come from an Other, he says: "Instead, it is *systemic*—that is, system-immanent—violence." Byung-Chul Han, *The Burnout Society*, Stanford University Press, 2015.

3. S. Peter Warren, "On Self-Licking Ice Cream Cones," in *Cool Stars, Stellar Systems, and the Sun: Proceedings of the 7th Cambridge Workshop*, ASP Conference Series, vol. 26.

4. From President John F. Kennedy's speech at Rice University, September 12, 1962. John F. Kennedy Presidential Library and Museum archives. https://www.jfklibrary.org/archives/other-resources/john-f-kennedy-speeches/rice-university-19620912.

5. Abraham M. Nussbaum, *The Finest Traditions of My Calling*, Yale University Press, 2017, 254.

6. Maria Montessori, *The Secret of Childhood*, Ballantine Books, 1982.

7. Maria Montessori et al., *The Secret of Childhood* (Vol. 22 of the Montessori Series), 119, Montessori-Pierson Publishing Company, 2007. Translations of Montessori's account vary.

8. Marc Andreessen, "It's Time to Build," Andreessen Horowitz.

9. The first quote comes from Maria Montessori, *The Montessori Method*, trans. Anne Everett George, Frederick A. Stokes Company, 1912. This second quote can be found in another edition of *The Montessori Method*, 41, trans. Anne E. George, CreateSpace Independent Publishing Platform, 2008. Today we might say "adult," but Montessori wrote this well over fifty years ago at a time when "man" would have been understood inclusively.

10. *The Montessori Method*, 2008 edition, 92. I also recommend Suzanne Ross, co-founder of the Raven Foundation, for an exploration of the role of mimetic theory in Montessori education. In one excellent essay, she writes: "During a presentation, the mode of interaction is implicitly one of imitation. The teacher is openly modeling a focused engagement with the material for the child to imitate. Then the teacher withdraws, allowing the child to take her place. The moment of withdrawal is made possible because the teacher's attention toward the material has been absorbed or internalized by the child. The admiration the teacher modeled

for the object is now also the child's. The object has been brought into the child's view, but rather than become rivals for its possession, teacher and child share it freely. It is the child's open imitation and the teacher's respect for it that makes the withdrawn mediation possible." Suzanne Ross, "The Montessori Method: The Development of a Healthy Pattern of Desire in Early Childhood," *Contagion: Journal of Violence, Mimesis and Culture* 19, 2012.

11. This was my key takeaway from a conversation with Louis Kim, a vice president at Hewlett-Packard, when I talked to him in 2019 about the effect of mimesis in larger companies. Having never worked for long in a large company myself, I've spent the last several years talking to as many people as I can about how mimesis manifests itself in more traditional corporate structures.

12. We saw in Chapter 2 how mimetic forces created a reality distortion field for Steve Jobs, and how mimetic desire bends the truth for most of us in our everyday lives. The tendency of mimetic desire to obscure or bend the truth has negative effects for us as individuals, and those effects are multiplied within organizations (as we saw for Zappos and the Downtown Project in Chapter 3).

13. CBS News, August 14, 2008. https://gigaom.com/2008/08/14/419-interview-blockbuster-ceo-dazed-and-confused-but-confident-of-physicals/.

14. Austin Carr, "Is a Brash Management Style Behind Blockbuster's $65.4M Quarterly Loss?," *Fast Company*, May 2010.

15. See the Benedictine idea of *habitare secum*, "to dwell with oneself," for further exploration. Nobody is totally *alone* in silence. We are still in relationship with others—they simply aren't present. Through silence, we peel back those relationships that hinder us and become reacquainted with those who help us live out the fullness of our humanity.

16. Zachary Sexton, "Burn the Boats," *Medium*, August 12, 2014.

17. For a short introduction to the concept, see Steve Blank, "Why the Lean Start-Up Changes Everything," *Harvard Business Review*, May 2013.

18. Eric Ries, "Minimum Viable Product: A Guide," *Startup Lessons Learned* (blog), August 3, 2009. http://www.startuplessonslearned.com/2009/08/minimum-viable-product-guide.html.

19. Interview of Toni Morrison by Kathy Neustadt, "Writing, Editing, and Teaching," *Alumnae Bulletin of Bryn Mawr College*, Spring 1980.

20. Sam Walker, "Elon Musk and the Dying Art of the Big Bet," *Wall Street Journal*, November 30, 2019.

21. Peter J. Boettke and Frédéric E. Sautet, "The Genius of Mises and the Brilliance of Kirzner," GMU Working Paper in Economics No. 11–05, February 1, 2011.

22. Artificial intelligence can augment entrepreneurs the same way that AI is augmenting farming in many parts of the world: much as it controls temperature, water usage, and harvest conditions and accuracy on AI-powered farms, it may control server usage, inventory management, and hiring decisions in companies. But AI can only support an immanent style of entrepreneurship—*not* one in which a person undertakes the uniquely human task of entrepreneurial awareness and creation.

Chapter 8: The Mimetic Future

1. Christianna Reedy, "Kurzweil Claims That the Singularity Will Happen by 2045," *Futurism*, October 5, 2017.

2. Ian Pearson, "The Future of Sex Report: The Rise of the Robosexuals," *Bondara*, September 2015.

3. The 2007 AI-animated film *Beowulf* was called "creepy" because the animations looked too humanlike. Subsequently, the studios made them appear less human.

4. Heather Long, "Where Are All the Startups? U.S. Entrepreneurship Near 40-Year Low," *CNN Business*, September 2016.

5. Drew Desilver, "For Most U.S. Workers, Real Wages Have Barely Budged in Decades," Pew Research Center, August 2018.

6. I have drawn the word "disenchanted" from the work of the Canadian philosopher Charles Taylor and, before him, Max Weber and Friedrich Schiller.

7. In the Roman Catholic Church, the start of this period is usually tied to the Second Vatican Council, which ended in 1965. Hundreds of thousands of priests and religious brothers and sisters left their vows behind and millions more laity shrugged their shoulders. The same thing has happened in almost every mainlline Protestant denomination and in non-Christian religions worldwide—with the notable exception of Islam.

8. Scott Galloway, *The Four: The Hidden DNA of Amazon, Apple, Facebook, and Google*, Portfolio/Penguin, 2017. This is my summary of his main points.

9. Eric Johnson, "Google Is God, Facebook Is Love and Uber Is 'Frat Rock,' Says Brand Strategy Expert Scott Galloway," *Vox*, June 2017. "Google is God. I think it's replaced God for us. As societies become more wealthy, more educated, religious institutions tend to play a smaller role in their lives, yet our modern-day anxieties and questions grow. There's an enormous spiritual void for a divine intervention. . . . One in five queries posed to Google have never been asked before in the history of humankind. Think of a cleric, a rabbi, a priest, a teacher, a coach that has so much credibility that one in five questions posed to that individual have never been asked before."

10. Ross Douthat, *The Decadent Society: How We Became the Victims of Our Own Success*, 5, Avid Reader Press, 2020.

11. Douthat, *The Decadent Society*, 136.

12. As I write this in mid-2020, I could also use the example of one who denies the reality of COVID-19 disease transmission yet wonders why people continue to die.

13. Alexis de Tocqueville, *Democracy in America*, 644, trans., ed., and with an introduction by Harvey C. Mansfield and Delba Winthrop, University of Chicago Press, 2000. The quoted passage is in vol. 2, part 4, chapter 3 ("Sentiments are in accord with ideas to concentrate power") and concerns, as the title suggests, the centralization of power.

14. These imaginary differences will be products of our *misrecognition*, from the reality distortion that mimetic desire causes. Misrecognition is what causes groups to mistakenly see scapegoats as grotesque and dangerous.

15. The King James Bible, Proverbs 29:18 ("Where there is no vision, the people perish . . .").

16. Shoshana Zuboff, *The Age of Surveillance Capitalism: The Fight for a Human Future at the New Frontier of Power*, PublicAffairs, 2020.

17. In *The Age of Surveillance Capitalism*, Zuboff gives the following complete definition in the front matter of the book: "*Sur-veil-lance Cap-i-tal-ism, n*. 1. A new economic order that claims human experience as free raw material for hidden commercial practices of extraction, prediction, and sales; 2. A parasitic economic logic in which the production of goods and services is subordinated to a new global architecture of behavioral modification; 3. A rogue mutation of capitalism marked by concentrations of wealth, knowledge, and power unprec-

edented in human history; 4. The foundational framework of a surveillance economy; 5. As significant a threat to human nature in the twenty-first century as industrial capitalism was to the natural world in the nineteenth and twentieth; 6. The origin of a new instrumentarian power that asserts dominance over society and presents startling challenges to market democracy; 7. A movement that aims to impose a new collective order based on total certainty; 8. An expropriation of critical human rights that is best understood as a coup from above: an overthrow of the people's sovereignty."

18. Matt Rosoff, "Here's What Larry Page Said on Today's Earnings Call," *Business Insider*, October 13, 2011.

19. George Gilder, *Life After Google: The Fall of Big Data and the Rise of the Blockchain Economy*, 21, Regnery Gateway, 2018.

20. Catriona Kelly and Vadim Volkov, "Directed Desires: Kul'turnost' and Consumption," in *Constructing Russian Culture in the Age of Revolution 1881–1940*, Oxford University Press, 1998.

21. Nanci Adler, *Keeping Faith with the Party: Communist Believers Return from the Gulag*, 20, Indiana University Press, 2012.

22. Nanci Adler, "Enduring Repression: Narratives of Loyalty to the Party Before, During and After the Gulag," *Europe-Asia Studies* 62, no. 2, 2010, 211–34.

23. Langdon Gilkey, *Shantung Compound: The Story of Men and Women Under Pressure*, 108, HarperOne, 1966.

24. Girard, *The One by Whom Scandal Comes*, 74.

25. This novel is explained beautifully by author and Nobel laureate Czesław Miłosz in his 1953 nonfiction book *The Captive Mind*, Secker and Warburg, 1953.

26. The idea of a pilled shortcut is a common trope in literature and film: soma in *Brave New World* and blue pills in *The Matrix* stand out as more widely known examples.

27. Martin Heidegger, *Discourse on Thinking: A Translation of Gelassenheit*, Harper & Row, 1966.

28. Iain McGilchrist, *The Master and His Emissary: The Divided Brain and the Making of the Western World*, Yale University Press, 2019.

29. A company "culture" is a set of things that a people considers sacred, whether in a country or a company. The word comes from the Latin *cultus*. Culture can't be fully understood without an understanding of *cult*, or religious ritual. Engineering a culture is engineering a religion.

30. René Girard and Chantre Benoît, *Battling to the End: Conversations with Benoît Chantre*, Michigan State University Press, 2009.

31. Daniel Kahneman's *Thinking, Fast and Slow* (Farrar, Straus and Giroux, 2015) does *not* correspond to calculating thought and meditative thought. Thinking fast and thinking slow are both forms of calculating thought, just different forms and speeds of it.

32. Virginia Hughes, "How the Blind Dream," *National Geographic*, February 2014.

33. Michael Polanyi and Mary Jo Nye, *Personal Knowledge: Towards a Post-Critical Philosophy*, University of Chicago Press, 2015.

34. Full disclosure: my wife, Claire, was the company's first employee in 2018 and has since become the director of business development.

35. By "invention," I don't mean that it was invented by any one person or group of people. The modern market economy, like the scapegoat mechanism, was a development that transcended any explicit engineering—it happened organically, as people wanted to find better ways to exchange goods with one another. The Girardian scholars Jean-Pierre Dupuy and Paul Dumouchel have both contributed greatly to the discussion of modern economics and

mimetic theory. My presentation of the role of the market economy here as the "second in-vention" is my own, and the result of a synthesis of a lot of ideas that I have derived in large part from these thinkers.

36. "Naval Ravikant—The Person I Call Most for Startup Advice," episode 97, *The Tim Ferriss Show* podcast, August 18, 2015.

37. Episode 1309 of *The Joe Rogan Experience* podcast, June 5, 2019.

38. Annie Dillard, *The Abundance: Narrative Essays Old and New*, 36, Ecco, 2016.

39. Dillard, *The Abundance*, 36.

40. Ibid.

Afterword

1. Interview with James G. Williams, "Anthropology of the Cross," 283–86, *The Girard Reader*, ed. James G. Williams, Crossroad, 1996.

2. Stephen King, *On Writing: A Memoir of the Craft*, 77, Scribner, 2000.

3. Cynthia Haven, "René Girard: Stanford's Provocative Immortel Is a One-Man Institution," *Stanford News*, June 11, 2008.

Appendix A: Glossary

1. Olivia Solon, "Richard Dawkins on the Internet's Hijacking of the Word 'Meme,'" *Wired*, June 2013.

SOURCES

Ackerman, Andy, dir. "The Parking Space." *Seinfeld*. DVD. New York: Castle Rock Entertain-ment, 1992.

———, dir. "The Soul Mate." *Seinfeld*. DVD. New York: Castle Rock Entertainment, 1996.

Adler, Nanci. "Enduring Repression: Narratives of Loyalty to the Party Before, During, and After the Gulag." *Europe-Asia Studies* 62, no. 2 (2010): 211–34. https://doi.org/10.1080/09668130903506797.

Agonie des Eros. Berlin: Matthes und Seitz, 2012.

Alberg, Jeremiah. *Beneath the Veil of the Strange Verses: Reading Scandalous Texts*. East Lansing: Michigan State University Press, 2013.

Alison, James. *The Joy of Being Wrong: Original Sin Through Easter Eyes*. New York: Crossroad, 2014.

———. *The Palgrave Handbook of Mimetic Theory and Religion*. Edited by Wolfgang Palaver. New York: Palgrave Macmillan, 2017.

Anspach, Mark Rogin, ed. *The Oedipus Casebook: Reading Sophocles' Oedipus the King*. Trans-lated by William Blake Tyrrell. East Lansing: Michigan State University Press, 2019.

———. *Vengeance in Reverse: The Tangled Loops of Violence, Myth, and Madness*. East Lansing: Michigan State University Press, 2017.

Antonello, Pierpaolo, and Paul Gifford, eds. *Can We Survive Our Origins? Readings in René*

Girard's Theory of Violence and the Sacred. East Lansing: Michigan State University Press, 2015.

———, eds. *How We Became Human: Mimetic Theory and the Science of Evolutionary Origins.* East Lansing: Michigan State University Press, 2015.

Antonello, Pierpaolo, and Heather Webb. *Mimesis, Desire, and the Novel: René Girard and Literary Criticism.* East Lansing: Michigan State University Press, 2015.

Ariely, Dan. *The (Honest) Truth About Dishonesty: How We Lie to Everyone—Especially Ourselves.* New York: Harper Perennial, 2013.

Astell, Ann W. "Saintly Mimesis, Contagion, and Empathy in the Thought of René Girard, Edith Stein, and Simone Weil." *Shofar* 22, no. 2 (2004): 116–31. Accessed May 10, 2020. https://www.jstor.org/stable/42943639.

Auerbach, Erich. *Mimesis: The Representation of Reality in Western Literature.* Translated by Willard R. Trask. Ewing, NJ: Princeton University Press, 2013.

Bahcall, Safi. *Loonshots: How to Nurture the Crazy Ideas That Win Wars, Cure Diseases, and Transform Industries.* New York: St. Martin's, 2019.

Bailie, Gil. *Violence Unveiled: Humanity at the Crossroads.* New York: Crossroad, 2004.

Balter, Michael. "Strongest Evidence of Animal Culture Seen in Monkeys and Whales." *Science Magazine,* April 25, 2013.

Bandera, Cesáreo. *The Humble Story of Don Quixote: Reflections on the Birth of the Modern Novel.* Washington, DC: Catholic University of America Press, 2006.

———. *A Refuge of Lies: Reflections on Faith and Fiction.* East Lansing: Michigan State University Press, 2013.

Barragan, Rodolfo Cortes, Rechele Brooks, and Andrew N. Meltzoff. "Altruistic Food Sharing Behavior by Human Infants After a Hunger Manipulation." *Scientific Reports* 10, no. 1 (2020). https://doi.org/10.1038/s41598-020-58645-9.

Bateson, Gregory. *Steps to an Ecology of Mind.* Chicago: University of Chicago Press, 2000.

Bergreen, Laurence. *Over the Edge of the World: Magellan's Terrifying Circumnavigation of the Globe.* New York: Perennial, 2004.

Berry, Steven E., and Michael Hardin. *Reading the Bible with René Girard: Conversations with Steven E. Berry.* Lancaster, PA: JDL, 2015.

Borch, Christian. *Social Avalanche: Crowds, Cities and Financial Markets.* Cambridge, UK: Cambridge University Press, 2020.

Bubbio, Paolo Diego. *Intellectual Sacrifice and Other Mimetic Paradoxes.* East Lansing: Michigan State University Press, 2018.

Buckenmeyer, Robert G. *The Philosophy of Maria Montessori: What It Means to Be Human.* Bloomington, IN: Xlibris, 2009.

Burgis, Luke, and Joshua Miller. *Unrepeatable: Cultivating the Unique Calling of Every Person.* Steubenville, OH: Emmaus Road, 2018.

Burkert, Walter, René Girard, and Jonathan Z. Smith. *Violent Origins: Walter Burkert, René Girard, and Jonathan Z. Smith on Ritual Killing and Cultural Formation.* Edited by Robert G. Hamerton-Kelly. Stanford, CA: Stanford University Press, 1988.

Buss, David M. *The Evolution of Desire: Strategies of Human Mating.* New York: Basic Books, 2016.

Canetti, Elias. *Crowds and Power.* Translated by Carol Stewart. New York: Farrar, Straus and Giroux, 1984.

Card, Orson Scott. *Unaccompanied Sonata and Other Stories.* New York: Dial, 1981.

Carse, James P. *Finite and Infinite Games*. New York: Free Press, 2013.

Cayley, David, ed. *The Ideas of René Girard: An Anthropology of Religion and Violence*. Independently published, 2019.

"The Century of the Self." *The Century of the Self*. London: BBC Two, March 2002.

Chelminski, Rudolph. *The Perfectionist: Life and Death in Haute Cuisine*. New York: Gotham Books, 2006.

Cialdini, Robert B. *Pre-suasion: A Revolutionary Way to Influence and Persuade*. New York: Simon and Schuster Paperbacks, 2018.

Collins, Brian. *Hindu Mythology and the Critique of Sacrifice: The Head Beneath the Altar*. Delhi: Motilal Banarasidas, 2018.

Collins, James C. *Good to Great*. New York: Harper Business, 2001.

Cowdell, Scott. *René Girard and the Nonviolent God*. Notre Dame, IN: University of Notre Dame Press, 2018.

———. *René Girard and Secular Modernity: Christ, Culture, and Crisis*. Notre Dame, IN: University of Notre Dame Press, 2015.

Cowen, Tyler. *What Price Fame?* Cambridge, MA: Harvard University Press, 2000.

Coyle, Daniel. *The Culture Code: The Secrets of Highly Successful Groups*. Read by Will Damron. Newark, NJ: Audible, 2018. Audiobook.

Crawford, Matthew B. *Shop Class as Soulcraft: An Inquiry into the Value of Work*. New York: Penguin Books, 2010.

Csikszentmihalyi, Mihaly. *Flow: The Psychology of Optimal Experience*. New York: Harper Perennial Modern Classics, 2009.

Danco, Alex. "Secrets About People: A Short and Dangerous Introduction to René Girard." April 28, 2019. https://alexdanco.com/2019/04/28/secrets-about-people-a-short-and-dangerous-introduction-to-rené-girard/.

Davies, Simone, and Hiyoko Imai. *The Montessori Toddler: A Parent's Guide to Raising a Curious and Responsible Human Being*. New York: Workman, 2019.

Dawson, David. *Flesh Becomes Word: A Lexicography of the Scapegoat or, the History of an Idea*. East Lansing: Michigan State University Press, 2013.

Deleuze, Gilles, and Félix Guattari. *Anti-Oedipus: Capitalism and Schizophrenia*. New York: Penguin Books, 2009.

DiMaggio, Paul J., and Walter W. Powell. "The Iron Cage Revisited: Institutional Isomorphism and Collective Rationality in Organizational Fields." *American Sociological Review* 48, no. 2 (1983): 147–60. Accessed February 21, 2020. https://www.jstor.org/stable/2095101.

Doerr, John E. *Measure What Matters: How Google, Bono, and the Gates Foundation Rock the World with OKRs*. New York: Portfolio/Penguin, 2018.

Douglas, Mary. *Purity and Danger: An Analysis of Concepts of Pollution and Taboo, with a New Preface by the Author*. Vol. 93. London: Routledge, 2002.

Dumouchel, Paul. *The Ambivalence of Scarcity and Other Essays*. East Lansing: Michigan State University Press, 2014.

———. *Barren Sacrifice: An Essay on Political Violence*. Translated by Mary Baker. East Lansing: Michigan State University Press, 2015.

Dumouchel, Paul, Luisa Damiano, and Malcolm DeBevoise. *Living with Robots*. Cambridge, MA: Harvard University Press, 2017.

Dupuy, Jean-Pierre. *Economy and the Future: A Crisis of Faith*. Translated by Malcolm B. DeBevoise. East Lansing: Michigan State University Press, 2014.

———, ed. *Self-Deception and Paradoxes of Rationality*. Stanford, CA: CSLI, 1997.

———. *A Short Treatise on the Metaphysics of Tsunamis*. Translated by Malcolm DeBevoise. East Lansing: Michigan State University Press, 2015.

Durkheim, Émile. *The Elementary Forms of Religious Life*. Edited by Mark Sydney Cladis. Translated by Carol Cosman. Oxford: Oxford University Press, 2008.

Epstein, Mark. *Open to Desire: The Truth About What the Buddha Taught*. New York: Gotham Books, 2006.

Erwin, Michael S. *Lead Yourself First—Inspiring Leadership Through Solitude*. New York: Bloomsbury, 2017.

Fan, Rui, Jichang Zhao, Yan Chen, and Ke Xu. "Anger Is More Influential Than Joy: Sentiment Correlation in Weibo." *PLoS ONE* 9, no. 10 (2014): e110184. https://doi.org/10.1371/journal.pone.0110184.

Farmer, Harry, Anna Ciaunica, and Antonia F. De C. Hamilton. "The Functions of Imitative Behaviour in Humans." *Mind and Language* 33, no. 4 (2018): 378–96. https://doi.org/10.1111/mila.12189.

Farneti, Roberto. *Mimetic Politics: Dyadic Patterns in Global Politics*. East Lansing: Michigan State University Press, 2015.

Fornari, Giuseppe. *A God Torn to Pieces: The Nietzsche Case*. Translated by Keith Buck. East Lansing: Michigan State University Press, 2013.

Fukuyama, Francis. *The End of History and the Last Man*. New York: Free Press, 2006.

Fullbrook, Edward, ed. *Intersubjectivity in Economics: Agents and Structures*. New York: Routledge, 2002.

Gardner, Stephen L. *Myths of Freedom: Equality, Modern Thought, and Philosophical Radicalism*. Westport, CT: Praeger, 1998.

Garrels, Scott R. *Mimesis and Science: Empirical Research on Imitation and the Mimetic Theory of Culture and Religion*. East Lansing: Michigan State University Press, 2011.

Germany, Robert. *Mimetic Contagion: Art and Artifice in Terence's "Eunuch."* New York: Oxford University Press, 2016.

Gifford, Paul. *Towards Reconciliation: Understanding Violence and the Sacred after René Girard*. Cambridge, UK: James Clarke, 2020.

Gilkey, Langdon. *Shantung Compound: The Story of Men and Women Under Pressure*. New York: HarperOne, 1975.

Girard, René. *Anorexia and Mimetic Desire*. Lansing: Michigan State University Press, 2013.

———. *Conversations with René Girard: Prophet of Envy*. Edited by Cynthia L. Haven. London: Bloomsbury Academic, 2020.

———. *Deceit, Desire, and the Novel*. Translated by Yvonne Freccero. Baltimore: Johns Hopkins University Press, 1976.

———. *Evolution and Conversion: Dialogues on the Origins of Culture*. London: Bloomsbury, 2017.

———. *The Girard Reader*. Edited by James G. Williams. New York: Crossroad Herder, 1996.

———. *I See Satan Fall Like Lightning*. New York: Orbis Books, 2001.

———. *Job: The Victim of His People*. Translated by Yvonne Freccero. Stanford, CA: Stanford University Press, 1987.

———. *Mimesis and Theory: Essays on Literature and Criticism, 1953–2005*. Edited by Robert Doran. Stanford, CA: Stanford University Press, 2011.

———. *Oedipus Unbound: Selected Writings on Rivalry and Desire*. Edited by Mark R. Anspach. Stanford, CA: Stanford University Press, 2004.

———. *The One by Whom Scandal Comes*. Translated by M. B. DeBevoise. East Lansing: Michigan State University Press, 2014.

———. *Resurrection from the Underground: Feodor Dostoevsky*. Edited and Translated by James G. Williams. East Lansing: Michigan State University Press, 2012.

———. *Sacrifice: Breakthroughs in Mimetic Theory*. Translated by Matthew Pattillo and David Dawson. East Lansing: Michigan State University Press, 2011.

———. *The Scapegoat*. Translated by Yvonne Freccero. Baltimore: Johns Hopkins University Press, 1989.

———. *A Theater of Envy: William Shakespeare*. South Bend, IN: St. Augustine's Press, 2004.

———. *To Double Business Bound: Essays on Literature, Mimesis, and Anthropology*. Baltimore: Johns Hopkins University Press, 1988.

———. *Violence and the Sacred*. Translated by Patrick Gregory. Baltimore: Johns Hopkins University Press, 1979.

———. *When These Things Begin: Conversations with Michel Treguer*. Translated by Trevor Cribben Merrill. East Lansing: Michigan State University Press, 2014.

Girard, René, and Benoît Chantre. *Battling to the End: Conversations with Benoît Chantre*. East Lansing: Michigan State University Press, 2009.

Girard, René, Robert Pogue Harrison, and Cynthia Haven. "Shakespeare: Mimesis and Desire." *Standpoint*, March 12, 2018.

Girard, René, Jean-Michel Oughourlian, and Guy Lefort. *Things Hidden Since the Foundation of the World*. Stanford, CA: Stanford University Press, 1987.

Girard, René, and Raymund Schwager. *René Girard and Raymund Schwager: Correspondence 1974–1991*. Edited by Joel Hodge, Chris Fleming, Scott Cowdell, and Mathias Moosbrugger. Translated by Chris Fleming and Sheelah Treflé Hidden. New York: Bloomsbury Academic, 2016.

Glaeser, Edward L. *Triumph of the City: How Our Greatest Invention Makes Us Richer, Smarter, Greener, Healthier, and Happier*. New York: Penguin Books, 2012.

Goffman, Erving. *The Presentation of Self in Everyday Life*. New York: Anchor Books, 1959.

Goodhart, Sandor. "In Tribute: René Girard, 1923–2015." *Religious Studies News*, December 21, 2015.

———. *The Prophetic Law: Essays in Judaism, Girardianism, Literary Studies, and the Ethical*. East Lansing: Michigan State University Press, 2014.

Goodhart, Sandor, Jørgen Jørgensen, Tom Ryba, and James Williams, eds. *For René Girard: Essays in Friendship and in Truth*. East Lansing: Michigan State University Press, 2010.

Grande, Per Bjørnar. *Desire: Flaubert, Proust, Fitzgerald, Miller, Lana Del Rey*. East Lansing: Michigan State University Press, 2020.

Granovetter, Mark S. *Society and Economy: Framework and Principles*. Cambridge, MA: Belknap Press of Harvard University Press, 2017.

Grant, Adam. *Give and Take: Why Helping Others Drives Our Success*. New York: Penguin Books, 2013.

———. *Originals: How Non-conformists Move the World*. New York: Penguin Books, 2017.

Greene, Robert. *The 48 Laws of Power*. New York: Penguin Books, 2000.

Greenfieldboyce, Nell. "Babies May Pick Up Language Cues in Womb." *NPR Morning Edition*, November 6, 2009.

Grote, Jim, and John McGeeney. *Clever as Serpents: Business Ethics and Office Politics*. Collegeville, MN: Liturgical, 1997.

Hamerton-Kelly, Robert. *Politics and Apocalypse*. East Lansing: Michigan University Press, 2008.

Han, Byung-Chul. *Abwesen: Zur Kultur und Philosophie des Fernen Ostens*. Berlin: Merve, 2007.

———. *Bitte Augen schließen: Auf der Suche nach einer anderen Zeit*. Berlin: Matthes und Seitz, 2013. E-book.

———. *Martin Heidegger*. Stuttgart: UTB, 1999.

———. *The Burnout Society*. Stanford, CA: Stanford Briefs, an imprint of Stanford University Press, 2015.

Hanna, Elizabeth, and Andrew N. Meltzoff. "Peer Imitation by Toddlers in Laboratory, Home, and Day-Care Contexts: Implications for Social Learning and Memory." *Developmental Psychology* 29, no. 4 (1993): 701–10. https://doi.org/10.1037/0012-1649.29.4.701.

Harari, Yuval Noah. *Homo Deus: A Brief History of Tomorrow*. London: Vintage, 2017.

Hardach, Sophie. "Do Babies Cry in Different Languages?," *New York Times*, November 14, 2019.

Haven, Cynthia. "René Girard: Stanford's Provocative Immortel Is a One-Man Institution." *Stanford News*, June 11, 2008.

Haven, Cynthia L. *Evolution of Desire: A Life of René Girard*. East Lansing: Michigan State University Press, 2018.

Heidegger, Martin. *Discourse on Thinking: A Translation of Gelassenheit*. New York: Harper & Row, 1966.

Henry, Todd, Rod Penner, Todd W. Hall, and Joshua Miller. *The Motivation Code: Discover the Hidden Forces That Drive Your Best Work*. New York: Penguin Random House, 2020.

Herman, Edward S., and Noam Chomsky. *Manufacturing Consent: The Political Economy of the Mass Media*. New York: Pantheon Books, 2002.

Hickok, Gregory. *The Myth of Mirror Neurons: The Real Neuroscience of Communication and Cognition*. New York: W. W. Norton, 2014.

Higgins, Tim. "Elon Musk's Defiance in the Time of Coronavirus." *Wall Street Journal*, March 20, 2020.

Hobart, Byrne, and Tobias Huber. "Manias and Mimesis: Applying René Girard's Mimetic Theory to Financial Bubbles." *SSRN Electronic Journal*, October 11, 2019. https://doi.org/10.2139/ssrn.3469465.

Holland, Tom. *Dominion: The Making of the Western Mind*. London: Little, Brown, 2019.

Hsieh, Tony. *Delivering Happiness: A Path to Profits, Passion, and Purpose*. New York: Grand Central, 2013.

Hughes, Virginia. "How the Blind Dream." *National Geographic*, February 2014.

Iacoboni, Marco. *Mirroring People: The Science of Empathy and How We Connect with Others*. New York: Picador, 2009.

Irvine, William Braxton. *On Desire: Why We Want What We Want*. New York: Oxford University Press, 2005.

Isaacson, Walter. *Steve Jobs*. New York: Simon & Schuster, 2011.

Jaffe, Eric. "Mirror Neurons: How We Reflect on Behavior." *Association for Psychological Science*, May 1, 2007.

Kahneman, Daniel, and Amos Tversky. "Prospect Theory: An Analysis of Decision Under Risk." *Econometrica* 47, no. 2 (1979): 263–91. Accessed September 16, 2020. https://doi.org/10.2307/1914185.

Kantor, Jodi, and Megan Twohey. *She Said: Breaking the Sexual Harassment Story That Helped Ignite a Movement*. New York: Penguin Books, 2019.

Kaplan, Grant. *René Girard, Unlikely Apologist: Mimetic Theory and Fundamental Theology*. Notre Dame, IN: University of Notre Dame Press, 2016.

Karniouchina, Ekaterina V., William L. Moore, and Kevin J. Cooney. "Impact of 'Mad Money' Stock Recommendations: Merging Financial and Marketing Perspectives." *Journal of Marketing* 73, no. 6 (2009): 244–66. Accessed February 12, 2020. https://www.jstor.org/stable /20619072.

Kethledge, Raymond Michael, and Michael S. Erwin. *Lead Yourself First: Inspiring Leadership Through Solitude.* London: Bloomsbury, 2019.

King, Stephen. *On Writing: A Memoir of the Craft.* New York: Scribner, 2010.

———. "Stephen King: How I Wrote Carrie." *Guardian*, April 4, 2014.

Kirwan, Michael. *Discovering Girard.* Cambridge, MA: Cowley, 2005.

Kirzner, Israel M. *Competition and Entrepreneurship.* Edited by Peter J. Boettke and Frédéric Sautet. Indianapolis: Liberty Fund, 2013.

Kofman, Fred, and Reid Hoffman. *The Meaning Revolution: The Power of Transcendent Leadership.* New York: Currency, 2018.

Kozinski, Thaddeus J. *Modernity as Apocalypse: Sacred Nihilism and the Counterfeits of Logos.* Brooklyn: Angelico, 2019.

Kramer, Rita. *Maria Montessori: A Biography.* New York: Diversion, 1988.

Kurczewski, Nick. "Lamborghini Supercars Exist Because of a 10-Lira Tractor Clutch." *Car and Driver*, 2018.

Laloux, Frederic. *Reinventing Organizations: An Illustrated Invitation to Join the Conversation on Next-Stage Organizations.* Brussels: Nelson Parker, 2016.

Lamborghini, Tonino. *Ferruccio Lamborghini: La Storia Ufficiale.* Argelato, Italy: Minerva, 2016.

Lawtoo, Nidesh. *Conrad's Shadow: Catastrophe, Mimesis, Theory.* East Lansing: Michigan State University Press, 2016.

———. *(New) Fascism: Contagion, Community, Myth.* East Lansing: Michigan State University Press, 2019.

Lebreton, Maël, Shadia Kawa, Baudouin Forgeot D'Arc, Jean Daunizeau, and Mathias Pessiglione. "Your Goal Is Mine: Unraveling Mimetic Desires in the Human Brain." *Journal of Neuroscience* 32, no. 21 (2012): 7146–57. https://www.jneurosci.org/content/32/21/7146.

Levy, David. *Love and Sex with Robots: The Evolution of Human-Robot Relationships.* London: Duckworth Overlook, 2009.

Lewis, C. S. "The Inner Ring." Memorial Lecture at King's College, University of London, 1944. https://www.lewissociety.org/innerring/.

Lewis, Mark. "Marco Pierre White on Why He's Back Behind the Stove for TV's Hell's Kitchen." *Caterer and Hotelkeeper*, April 25, 2007. https://www.thecaterer.com/news/restaurant /exclusive-marco-pierre-white-on-why-hes-back-behind-the-stove-for-tvs-hells-kitchen.

Lieberman, Daniel Z., and Michael E. Long. *The Molecule of More: How a Single Molecule in Your Brain Drives Love, Sex, and Creativity—and Will Determine the Fate of the Human Race.* Dallas: BenBella Books, 2018.

Lillard, Angeline Stoll. *Montessori: The Science Behind the Genius.* Oxford: Oxford University Press, 2008.

Lindsley, Art. "C. S. Lewis: Beware the Temptation of the 'Inner Ring.'" Institute for Faith, Work and Economics, May 2019.

Lippmann, Walter. *Public Opinion: A Classic in Political and Social Thought.* Charleston, SC: Feather Trail, 2010.

Lombardo, Nicholas E. *The Logic of Desire: Aquinas on Emotion.* Washington, DC: Catholic University of America Press, 2011.

Long, Heather. "Where Are All the Startups? U.S. Entrepreneurship Near 40-Year Low." *CNN Business*, September 8, 2016.

Lorenz, Konrad. *On Aggression*. New York: Houghton Mifflin Harcourt, 1974.

Lucas, Henry C. *The Search for Survival: Lessons from Disruptive Technologies*. Santa Barbara, CA: Praeger, 2012.

MacIntyre, Alasdair C. *After Virtue: A Study in Moral Theory*. Notre Dame, IN: University of Notre Dame Press, 2012.

Mampe, Birgit, Angela D. Friederici, Anne Christophe, and Kathleen Wermke. "Newborns' Cry Melody Is Shaped by Their Native Language." *Current Biology* 19, no. 23 (2009). https://doi .org/10.1016/j.cub.2009.09.064.

Mandelbrot, Benoit B., and Richard L. Hudson. *The Misbehavior of Markets: A Fractal View of Financial Turbulence*. New York: Basic Books, 2006.

Martino, Ernesto de. *The Land of Remorse: A Study of Southern Italian Tarantism*. Translated by Dorothy Louise Zinn. London: Free Association Books, 2005.

McCormack, W. J. *Enigmas of Sacrifice: A Critique of Joseph M. Plunkett and the Dublin Insurrection of 1916*. East Lansing: Michigan State University Press, 2016.

McGilchrist, Iain. *The Master and His Emissary: The Divided Brain and the Making of the Western World*. New Haven, CT: Yale University Press, 2019.

McKenna, Andrew. *Semeia 33: René Girard and Biblical Studies*. Atlanta: Society of Biblical Literature, 1985.

Medvedev, Roy Aleksandrovich, and George Shriver. *Let History Judge: The Origins and Consequences of Stalinism*. New York: Columbia University Press, 1989.

Meltzoff, Andrew N., and M. Keith Moore. "Newborn Infants Imitate Adult Facial Gestures." *Child Development* 54 (1983): 702–09. Photo credit: A. N. Meltzoff and M. K. Moore. *Science* 198 (1977): 75–78.

Meltzoff, Andrew. "Born to Learn: What Infants Learn from Watching Us." In *The Role of Early Experience in Infant Development*. Edited by Nathan A. Fox, Lewis A. Leavitt, John G. Warhol, 1–10. New Brunswick, NJ: Johnson & Johnson, 1999.

———. "The Human Infant as Homo Imitans." In *Social Learning: Psychological and Biological Perspectives*. Edited by Thomas R. Zentall and B. G. Galef Jr., 319–41. East Sussex, UK: Psychology Press, 1988.

———. "Imitation, Objects, Tools, and the Rudiments of Language in Human Ontogeny." *Human Evolution* 3, no. 1–2 (1988): 45–64. https://doi.org/10.1007/bf02436590.

———. "Like Me: A Foundation for Social Cognition." In *Developmental Science*, 126–34. Hoboken, NJ: Wiley-Blackwell, 2007. https://doi.org/10.1111/j.1467–7687.2007.00574.x.

———. "Origins of Social Cognition: Bidirectional Self-Other Mapping and the 'Like-Me' Hypothesis." In *Navigating the Social World: What Infants, Children, and Other Species Can Teach Us*. Edited by Mahzarin R. Banaji and Susan A. Gelman, 139–44. Oxford: Oxford University Press, 2013.

———. "Understanding the Intentions of Others: Re-enactment of Intended Acts by 18-Month-Old Children." *Developmental Psychology* 31, no. 5 (1995): 838–50. https://doi .org/10.1037/0012–1649.31.5.838.

Meltzoff, Andrew N., and Rechele Brooks. "Self-Experience as a Mechanism for Learning About Others: A Training Study in Social Cognition." *Developmental Psychology* 44, no. 5 (2008): 1257–65. https://doi.org/10.1037/a0012888.

Meltzoff, Andrew N., Patricia K. Kuhl, Javier Movellan, and Terrence J. Sejnowski. "Foundations for a New Science of Learning." *Science Magazine* 325, no. 5938 (July 17, 2009): 284–88. https://doi.org/10.1126/science.1175626.

Meltzoff, Andrew N., and Peter J. Marshall. "Human Infant Imitation as a Social Survival Circuit." *Current Opinion in Behavioral Sciences* 24 (2018): 130–36. https://doi.org/10.1016/j.cobeha.2018.09.006.

Meltzoff, Andrew N., Rey R. Ramírez, Joni N. Saby, Eric Larson, Samu Taulu, and Peter J. Marshall. "Infant Brain Responses to Felt and Observed Touch of Hands and Feet: An MEG Study." *Developmental Science* 21, no. 5 (2017). https://doi.org/10.1111/desc.12651.

Merrill, Trevor Cribben. *The Book of Imitation and Desire: Reading Milan Kundera with René Girard*. London: Bloomsbury, 2014.

Merton, Thomas. *New Seeds of Contemplation*. New York: New Directions Books, 2007.

———. *No Man Is an Island*. San Diego, CA: Harcourt, 1955.

———. *Thoughts in Solitude*. New York: Farrar, Straus and Giroux, 2011.

Montessori, Maria. *The Absorbent Mind: A Classic in Education and Child Development for Educators and Parents*. New York: Henry Holt, 1995.

———. *The Montessori Method*. Translated by Anne E. George. Scotts Valley, CA: CreateSpace Independent Publishing Platform, 2008.

———. *The Secret of Childhood*. New York: Ballantine Books, 1982.

Murphy, James Bernard. *A Genealogy of Violence and Religion: René Girard in Dialogue*. Chicago: Sussex Academic, 2018.

Nisbet, Robert A. *History of the Idea of Progress*. London: Routledge, 2017.

Noelle-Neumann, Elisabeth. *The Spiral of Silence: Public Opinion—Our Social Skin*. Chicago: University of Chicago Press, 1994.

Novak, Michael, et al. *Social Justice Isn't What You Think It Is*. New York: Encounter Books, 2015.

Nowrasteh, Cyrus, dir. *The Stoning of Soraya M.* Amazon. Paramount Home Entertainment, June 26, 2009. https://www.amazon.com/Stoning-Soraya-M-Shohreh-Aghdashloo/dp/B008Y79Z66/ref=sr_1_1?dchild=1&keywords=stoning+of+soraya+m.&qid=1585427690&sr=8-1.

Nuechterlein, Paul J. "René Girard: The Anthropology of the Cross as Alternative to Post-Modern Literary Criticism." *Girardian Lectionary*, October 2002.

Onaran, Yalman, and John Helyar. "Fuld Solicited Buffett Offer CEO Could Refuse as Lehman Fizzled." *Bloomberg*, 2008.

Ordóñez, Lisa, Maurice Schweitzer, Adam Galinsky, and Max Bazerman. "Goals Gone Wild: The Systematic Side Effects of Over-Prescribing Goal Setting." Harvard Business School, 2009.

Orléan, André. *The Empire of Value: A New Foundation for Economics*. Translated by M. B. DeBevoise. Cambridge, MA: MIT Press, 2014.

O'Shea, Andrew. *Selfhood and Sacrifice: René Girard and Charles Taylor on the Crisis of Modernity*. New York: Continuum International, 2010.

Oughourlian, Jean-Michel. *The Genesis of Desire*. East Lansing: Michigan State University Press, 2010.

———. *The Mimetic Brain*. Translated by Trevor Cribben Merrill. East Lansing: Michigan State University Press, 2016.

———. *Psychopolitics: Conversations with Trevor Cribben Merrill*. Translated by Trevor Cribben Merrill. East Lansing: Michigan State University Press, 2012.

Palaver, Wolfgang. *René Girard's Mimetic Theory*. East Lansing: Michigan State University Press, 2013.

Palaver, Wolfgang, and Richard Schenk, eds. *Mimetic Theory and World Religions*. East Lansing: Michigan State University Press, 2017.

Palmer, Parker J. *Let Your Life Speak: Listening for the Voice of Vocation*. San Francisco: Jossey-Bass, 2000.

Pearson, Ian. "The Future of Sex Report." *Bondara*, September 2015.

Pérez, Julián Carrón. *Disarming Beauty: Essays on Faith, Truth, and Freedom*. Notre Dame, IN: University of Notre Dame Press, 2017.

"Peter Thiel on René Girard." ImitatioVideo. YouTube. 2011. https://www.youtube.com/watch?v=esk7W9Jowtc.

Pinker, Steven. *The Better Angels of Our Nature: Why Violence Has Declined*. New York: Penguin, 2012.

Polanyi, Michael, and Mary Jo Nye. *Personal Knowledge: Towards a Post-critical Philosophy*. Chicago: University of Chicago Press, 2015.

Proust, Marcel. *In Search of Lost Time: The Captive, The Fugitive*. Modern Library Edition. Vol. 5. New York: Random House, 1993.

Qualls, Karl. Review of *Constructing Russian Culture in the Age of Revolution: 1881–1940*, edited by Catriona Kelly and David Shepherd. H-Russia, H-Net Reviews, February 2000. http://www.h-net.org/reviews/showrev.php?id=3813.

Reedy, Christianna. "Kurzweil Claims That the Singularity Will Happen by 2045." *Futurism*, October 5, 2015.

Reineke, Martha J. *Intimate Domain: Desire, Trauma, and Mimetic Theory*. East Lansing: Michigan State University Press, 2014.

———. *Sacrificed Lives: Kristeva on Women and Violence*. Bloomington: Indiana University Press, 1997.

Repacholi, Betty M., Andrew N. Meltzoff, Theresa M. Hennings, and Ashley L. Ruba. "Transfer of Social Learning Across Contexts: Exploring Infants' Attribution of Trait-Like Emotions to Adults." *Infancy* 21, no. 6 (2016): 785–806. https://doi.org/10.1111/infa.12136.

Repacholi, Betty M., Andrew N. Meltzoff, Tamara Spiewak Toub, and Ashley L. Ruba. "Infants' Generalizations About Other People's Emotions: Foundations for Trait-Like Attributions." *Developmental Psychology* 52, no. 3 (2016): 364–78. https://doi.org/10.1037/dev0000097.

Rocha, João Cezar de Castro. *Machado de Assis: Toward a Poetics of Emulation*. Translated by Flora Thomson-DeVeaux. East Lansing: Michigan State University Press, 2015.

———. *Shakespearean Cultures: Latin America and the Challenges of Mimesis in Non-hegemonic Circumstances*. Translated by Flora Thomson-DeVeaux. East Lansing: Michigan State University Press, 2019.

Rosenberg, Randall S. *The Givenness of Desire: Concrete Subjectivity and the Natural Desire to See God*. Toronto: University of Toronto Press, 2018.

Rosoff, Matt. "Here's What Larry Page Said on Today's Earnings Call." *Business Insider*, October 13, 2011.

Ross, Suzanne. "The Montessori Method: The Development of a Healthy Pattern of Desire in Early Childhood." *Contagion: Journal of Violence, Mimesis, and Culture* 19 (2012): 87–122. Accessed January 19, 2020. https://www.jstor.org/stable/41925335.

Ross, W. D. *Aristotle's Metaphysics. A Revised Text with Introduction and Commentary*. Oxford: Clarendon Press, 1924.

Sacks, David O., and Peter A. Thiel. *The Diversity Myth: Multiculturalism and Political Intolerance on Campus*. Oakland, CA: Independent Institute, 1998.

Sacks, Rabbi. "Introduction to Covenant and Conversation 5776 on Spirituality." https://rabbi sacks.org, October 7, 2015.

Schoeck, Helmut. *Envy: A Theory of Social Behaviour*. Indianapolis: Liberty, 1987.

Schulz, Bailey, and Richard Velotta. "Zappos CEO Tony Hsieh, Champion of Downtown Las Vegas, Retires." *Las Vegas Review-Journal*, August 24, 2020.

Scubla, Lucien. *Giving Life, Giving Death: Psychoanalysis, Anthropology, Philosophy*. Translated by Malcolm DeBevoise. East Lansing: Michigan State University Press, 2016.

Sexton, Zachary. "Burn the Boats." *Medium*, August 12, 2014.

Sheehan, George. *Running and Being: The Total Experience*. New York: Rodale, 2014.

Simonse, Simon. *Kings of Disaster: Dualism, Centralism and the Scapegoat King in Southeastern Sudan*. Kampala, Uganda: Fountain, 2017.

Sinek, Simon. *The Infinite Game*. New York: Penguin, 2019.

Singer, Isidore, and Cyrus Adler. *The Jewish Encyclopedia: A Descriptive Record of the History, Religion, Literature, and Customs of the Jewish People from the Earliest Times to the Present Day*. Vol. 10. Charleston, SC: Nabu, 2012.

Smee, Sebastian. *The Art of Rivalry: Four Friendships, Betrayals, and Breakthroughs in Modern Art*. New York: Random House, 2017.

Solon, Olivia. "Richard Dawkins on the Internet's Hijacking of the Word 'Meme.'" *Wired*, June 20, 2013.

Sorkin, Andrew Ross. *Too Big to Fail: The Inside Story of How Wall Street and Washington Fought to Save the Financial System—and Themselves*. New York: Penguin Books, 2018.

Soros, George. *The Alchemy of Finance*. Hoboken, NJ: Wiley, 2003.

———. "Fallibility, Reflexivity, and the Human Uncertainty Principle." *Journal of Economic Methodology* 20, no. 4 (January 13, 2014). https://doi.org/10.1080/1350178x.2013.859415.

Standing, E. M, and Lee Havis. *Maria Montessori: Her Life and Work*. New York: Plume, 1998.

Strenger, Carlo. *Critique of Global Unreason: Individuality and Meaning in the Twenty-First Century*. New York: Palgrave Macmillan, 2011.

———. *The Fear of Insignificance: Searching for Meaning in the Twenty-First Century*. New York: Palgrave Macmillan, 2016.

Subiaul, Francys. "What's Special About Human Imitation? A Comparison with Enculturated Apes." *Behavioral Sciences* 6, no. 3 (July 2016): 13. https://doi.org/10.3390/bs6030013.

Taleb, Nassim Nicholas. *Antifragile: Things That Gain from Disorder*. New York: Random House, 2014.

———. *Skin in the Game: Hidden Asymmetries in Daily Life*. New York: Random House, 2018.

Taylor, Charles. *A Secular Age*. Cambridge, MA: Belknap Press of Harvard University Press, 2018.

Thaler, Richard H. *Misbehaving: The Making of Behavioral Economics*. New York: W. W. Norton, 2016.

Thiel, Peter, and Blake Masters. *Zero to One: Notes on Startups, or How to Build the Future*. New York: Crown Business, 2014.

Thomson, Cameron, Sandor Goodhart, Nadia Delicata, Jon Pahl, Sue-Anne Hess, Glenn D. Smith, Eugene Webb, et al. *René Girard and Creative Reconciliation*. Edited by Thomas Ryba. Lanham, MD: Lexington Books, 2014.

THR Staff. "Fortnite, Twitch . . . Will Smith? 10 Digital Players Disrupting Traditional Hollywood." *HollyWood Reporter*, November 1, 2018.

Turkle, Sherry. *Alone Together: Why We Expect More from Technology and Less from Each Other.* New York: Basic Books, 2017.

Tversky, Amos, and Daniel Kahneman. "Rational Choice and the Framing of Decisions." *Journal of Business* 59, no. 4 (1986): S251–78. Accessed September 16, 2020. http://www.jstor.org /stable/2352759.

Tye, Larry. *The Father of Spin: Edward L. Bernays and the Birth of Public Relations.* New York: Henry Holt, 2002.

Tyrrell, William Blake. *The Sacrifice of Socrates: Athens, Plato, Girard.* East Lansing: Michigan State University Press, 2012.

Vattimo, Gianni, and René Girard. *Christianity, Truth, and Weakening Faith: A Dialogue.* Edited by Pierpaolo Antonello. Translated by William McCuaig. New York: Columbia University Press, 2010.

Von Hildebrand, Dietrich, and John F. Crosby. *Ethics.* Steubenville, OH: Hildebrand Project, 2020.

Wallace, David Foster. *Infinite Jest.* New York: Back Bay Books, 2016.

Waller, John. *The Dancing Plague: The Strange, True Story of an Extraordinary Illness.* Naperville, IL: Sourcebooks, 2009.

———. *A Time to Dance, a Time to Die: The Extraordinary Story of the Dancing Plague of 1518.* Duxford, UK: Icon Books, 2009.

Warren, James. *Compassion or Apocalypse: A Comprehensible Guide to the Thought of René Girard.* Washington, DC: Christian Alternative, 2013.

Warren, S. Peter. "On Self-Licking Ice Cream Cones." Paper presented at Cool Stars, Stellar Systems, and the Sun Seventh Cambridge Workshop, ASP Conference Series, vol. 26. 1992.

Weil, Simone, and Gustave Thibon. *Gravity and Grace.* Translated by Emma Crawford and Mario von der Ruhr. London/New York: Routledge, 2008.

Weinstein, Eric. "Interview with Peter Thiel." *The Portal.* Podcast audio. July 17, 2019.

Zuboff, Shoshana. *The Age of Surveillance Capitalism: The Fight for a Human Future at the New Frontier of Power.* New York: PublicAffairs, 2020.

INDEX

Note: Page numbers in *italics* refer to figures and photographs.